STELLARIS:
People of the Stars

BAEN BOOKS by Les Johnson

Mission to Methone

WITH BEN BOVA
Rescue Mode

WITH TRAVIS S. TAYLOR
Back to the Moon
On to the Asteroid

EDITED BY LES JOHNSON AND JACK MCDEVITT
Going Interstellar

EDITED BY LES JOHNSON AND ROBERT E. HAMPSON
Stellaris: People of the Stars

STELLARIS:
People of the Stars

Edited by
Les Johnson & Robert E. Hampson

"Foreword," © 2019 Robert E. Hampson; "Burn the Boats," © 2019 Sarah A. Hoyt; "Bridging," © 2019 William Ledbetter; "The Future of Intelligent Life in the Cosmos," © 2019 Martin Rees; "Stella Infantes," © 2019 Kacey Ezell and Philip Wohlrab; "Maintaining Crew Health and Mission Performance in Ventures Beyond Near-Earth Space," © 2019 Mark Shelhamer; "At the Bottom of the White," © 2019 Todd McCaffrey; "Pageants of Humanity," © 2019 Brent Roeder; *"Home Stellaris*—Working Track Report from the Tennessee Valley Interstellar Workshop," © 2019 Robert E. Hampson and Les Johnson; "Time Flies," © 2019 Kevin J. Anderson; "Our Worldship Broke!" © 2019 Jim Beall; "Nanny," © 2019 Les Johnson; "Those Left Behind," © 2019 Robert E. Hampson; "Securing the Stars: The Security Implications of Human Culture During Interstellar Flight," © 2019 Mike Massa; "The Smallest of Things," © 2019 Catherine L. Smith; "Biological and Medical Challenges of the Transition to *Homo Stellaris*," © 2019 Nikhil Rao, MD; "Exodus," © 2019 Daniel M. Hoyt; "Afterword," © 2019 Les Johnson; "Tennessee Valley Interstellar Workshop," © 2019 Joe Meany, Edward E. Montgomery, and John Preston

A Baen Books Original

Baen Publishing Enterprises
P.O. Box 1403
Riverdale, NY 10471
www.baen.com

ISBN: 978-1-4814-8425-1

Cover art by Sam Kennedy.
TVIW logo created by Debbie Hughes and used with the permission of the TVIW.

First printing, September 2019

Distributed by Simon & Schuster
1230 Avenue of the Americas
New York, NY 10020

Library of Congress Cataloging-in-Publication Data

Names: Johnson, Les (Charles Les), editor. | Hampson, Robert E., editor.
Title: Stellaris : people of the stars / edited by Les Johnson and Robert E. Hampson.
Description: Riverdale, NY : Baen, [2019]
Identifiers: LCCN 2019020168 | ISBN 9781481484251 (paperback)
Subjects: LCSH: Science fiction, American. | Interstellar travel—Fiction. |
 Interstellar travel. | Future, The. | Forecasting. | BISAC: FICTION /
 Science Fiction / High Tech. | FICTION / Science Fiction / Short Stories.
 | SCIENCE / Life Sciences / Evolution.
Classification: LCC PS648.S3 S74 2019 | DDC 813/.0876208—dc23 LC record available at
https://lccn.loc.gov/2019020168

10 9 8 7 6 5 4 3 2 1
Pages by Joy Freeman (www.pagesbyjoy.com)
Printed in the United States of America

To those who have served on the Board of Directors for the Tennessee Valley Interstellar Workshop (TVIW). These volunteer visionaries took an ill-formed idea and crafted it into one of the most successful private interstellar-focused space advocacy groups in the world: Martha Knowles, Ken Roy, John Preston, Robert Kennedy, David Fields, Yohon Lo, Doug Loss, Sandy Montgomery, Jim Moore, Marc Millis, Joe Meany, and Paul Gilster. Ad astra indeed!

Contents

To the dreamers . . .

Foreword

Becoming the People of the Stars

This volume has its origins in the March 2016 Tennessee Valley Interstellar Workshop (TVIW) Symposium held at the Chattanooga Choo Choo Hotel in Chattanooga, TN. The TVIW "working tracks" from each symposium provide an opportunity to have extended conversations about some of the "big issues" involved in making the transition to interstellar exploration and colonization. In 2016, one of the working tracks was entitled "Homo Stellaris" and the participants were charged with examining the transition of society, society's mindset, and the human body to a life among the stars. A more complete report from that event is included later in this book (see "*Homo Stellaris*—Working Track Report from the Tennessee Valley Interstellar Workshop 2016"). One of the more intriguing questions to come out of the working track was, "Will the interstellar explorers be *human* as we define it?" In other words, will the inevitable changes to our society—not to mention the necessary changes to physiology and psychology—change those explorers to something other than *Homo sapiens*? If so, how then will those explorers preserve their essential humanity, rather than simply becoming what amounts to biological robotic probes?

HUMANS IN SPACE

The original Apollo program was extremely lucky to lose only three astronauts to spacecraft complications (Apollo 1) and that incident occurred on the ground. The only (publicized) potential

1

loss-of-life incident in space was Apollo 13 and it was resolved successfully. On the other hand, two complete crews were lost to space shuttle accidents, and, each time, missions resumed only after extensive hand-wringing and finger-pointing.

If humans are to eventually go to the stars, they will first have to go out into space, beyond the ISS, beyond the Moon, even beyond Mars. To do that, people not only have to *want* to go, they have to do so in the face of risk and loss. There is hope, however, and that hope comes from science fiction (SF). The more society questions and thinks about not only the problems but also uses their imaginations to create solutions, the better prepared humans will be to adapt and overcome the risks of living, working, and *thriving* in space.

Early SF didn't worry too much about adapting humans to space or other planets—mainly because so little was known about the differences humans would encounter once they left the surface of Earth. Space was still "the ether" and either just like the atmosphere (only thinner) or simply ignored. Hence, we had images of airborne ships "sailing" to the Moon, astronauts riding on the outside of rocket ships, and giant cannons that simply fired projectiles to the Moon where they would be greeted by outlandish beings who breathed air and lived on human-style food. Once it was generally accepted that space was a vacuum, and that both space and other planetary environments held many hazards to human health, SF turned to the idea that humans would naturally take their environment with them and re-create it on other worlds.

One of the most common book-cover images from SF is the space-suited astronaut on a hill looking over a valley of domes. This graphic image fits most perceptions of initial planetary habitats. Given what we now know about the air, soil, and radiation conditions on the Moon, Mars, and Venus, those assumptions are generally correct. In fact, most extraterrestrial communities will need to be underground (or inside rocky asteroids) for maximum protection from radiation, and they will thus seek to become as self-sustaining as possible. Breathable air will probably be generated artificially, then filtered and refreshed by plants. Likewise, water will be recycled and perhaps refreshed by mining asteroids. These are habitats in which humans can live without adaptation, much as a spacesuit captures a small volume of terrestrial life support and keeps a human from being exposed to the hazards of space.

But would it be possible to adapt *humans* to their environment, instead of the *environment* to humans?

It is important to note that simply providing a sheltering terrestrial environment will not prevent humans from adapting to the novel aspects of their new habitats. For example, research has shown that low-gravity environments result in a reduction of bone and muscle mass as well as changes in vision and heart function. While these are short-term adaptations which rely on adaptive mechanisms and not *evolution* of the human form, would it be possible to *intentionally* alter humans for space?

SF has already been there, in stories ranging from simple genetic improvement of human health, to wholesale alteration of the human body into completely alien forms. While this level of gene editing is still well outside of current capabilities, the field of tissue engineering is rapidly developing, as shown by the recent announcements of lab-created simple human organs such as the bladder, human ears grown on the backs of laboratory mice ears, sprayable skin cells for burn repair, and efforts to "3-D print" liver and kidney cells.

What other types of gene engineering might be desirable to adapt humans to space? For this question, we need look no further than our own oceans. Fish and marine mammals provide examples of adaptation such as pressure tolerance, maneuverability in a fluid/weightless environment, temperature extremes, oxygen extremes, sulfur-dependent organisms, alterations of circadian rhythm, and independence from sleep. "Natural antifreeze" copied from Arctic cod may make the difference in adapting or engineering humans to cryogenic stasis during long-duration spaceflights. While we may never reach the state of total freedom to choose alternate bodies, many examples—not to mention source materials—for engineering humans for life in space are already right here on Earth.

At the same time, we must ensure that human society has the *will* to tolerate the extreme risks of long-duration space exploration. Our society has become decidedly risk-averse and has difficulty making (and funding) long-range and long-term investments in social projects and technology. For some, this is compounded by an attitude that "there are too many problems at home to waste money in space!" It is a common theme in SF that space is a frontier, and in many ways only a small subset of society will embrace the *necessity* as well as the attraction of that frontier. Space may very well be colonized by the misfits, by those for whom it is no longer possible to fit into Earthbound society, or by individuals who decide to balance significant risk with even greater reward. For these reasons, the societies we build in space may be totally unlike our current experience.

SHAPING THE FUTURE

What better way to explore these concepts than through SF? It is a playground, a sandbox in which we can experiment with ideas and concepts that are beyond current capabilities. Perhaps through fiction we can encourage people to think about these issues, and begin to make the adaptations and accommodations to lessen the shock that such changes will produce. As we consider whether (or how) to adapt humans to new environments or adapt those environments to better suit humans, we also need to examine the *science* behind human experiences in space. In this anthology we have combined nonfiction essays and SF short stories to examine the motivations, the hazards, and the adaptations that will be encountered as humans move into a permanent presence in space and become *Homo Stellaris*—**the People of the Stars**.

Robert E. Hampson, PhD
Kernersville, NC, October 2018

Burn the Boats

Sarah A. Hoyt

Sarah A. Hoyt has published over thirty novels (don't make her count!) in science fiction, fantasy, mystery, historical, and they-say-it's-romance. Also, over one hundred short stories (really don't make her count. We'll be here all day) in magazines like Analog and Asimov's, Weird Tales, and others, as well as a whole bunch of anthologies. However, since she broke into her aunt's house (really don't make her explain) at age six to watch the Moon landing, her first and last love has been science fiction. She's cleverly managed to guide one son into medicine and one into engineering, for the sole purpose of using them as sources to supply her own pitiful knowledge of the subjects. Thus armed, she hopes to be able to spend imaginary time in space, even if she'll never live there. Oh, yeah: She was born in Portugal, writes in her third language (if you ask her to say "moose and squirrel" you can't be her friend anymore), has won the Prometheus Award for her novel *DarkShip Thieves*, etc., etc. But mostly she's just happy to be writing science fiction.

They'd swept a path on the green-blue ice, so that it looked like a road, from the landing site to the village. On either side of it, the snowflakes crusted, gilded a pale orange by the light from Proxima Virginis.

Martha swallowed hard and bowed to the two waiting... men—she would have to remember to think of them as men—who stood on either side of the path as what? Guards? Escorts?

Their skin was too pale, they had no noses that she thought of as noses and they looked, for lack of a better term, slimy. She breathed through her mouth, so as not to detect what her nose insisted was a distinctly fish odor, and she squared her shoulders in her temperature-controlled suit. The men wore what looked like harem pants in a fabric that looked as if they'd skinned a fish and not cured the skin. And she had a feeling that was just a concession to the new arrivals, despite the cold making the skin of her own face go numb and her eyes sting.

They bowed to the survivors of Gloriana with a sort of fluid elegance that made all the alarm bells go off at the back of Martha's head. There was something here like the uncanny valley effect that had made androids a rare thing back on Earth because they looked just enough like men but weren't to set off subconscious alarm. Perhaps if Martha hadn't known these were humans, the same seed of Earth as her own people, this would be easier.

There wasn't much to the swept path. After half a mile, and cresting a rise that looked artificial, they came upon a village of clear igloos. They had obviously purified the water, making the bricks of the persistent blue-green algae. The igloos sparkled gold under the sun, but were too small to be habitations. Which was fine. The scouts and ambassadors who had arranged this deal for her people had told her they weren't habitations so much as covers over greenhouses, both to increase the concentration of oxygen and to keep the plants warm enough to grow. Below that were the actual houses, in ice caves, and below that still, the waters where these amphibian-adapted humans farmed the crabs and fish that provided the protein in their diets.

As Martha approached, followed by her people, she and they acting like polite refugees, each carrying only a small sack of possessions, not even sure what they would need in this new life, people flowed out of the igloos. Between the lower gravity of this much smaller world—Diana, which the people on Earth had called Ross 128 c—and the modifications that made it possible for them to spend so much of their time underwater, their chests were larger and it seemed to her their heads were smaller, though that might be an effect of the proportionally larger chests. Their arms seemed longer too, and spindly, and their hands had long inter-finger membranes. As for their legs, they seemed too long and too thin, like they bent in extra places.

Men, women and children were near-naked, though some wore skirts or pants that looked like fish skin, and most wore some kind of ornament, from earrings to what looked like tattoos.

They flowed out, in massed confusion behind a man up front, who surged forward with either an eager or threatening expression on his face as he undulated into a deep bow before her. "Welcome, Martha MacArthur," he said, "leader of our dead sister-world, Gloriana. Welcome to Diana, be welcome. And may your people and mine prosper together." His voice sounded too high and had weird echoes. Reports said these people saw much better in the murky water of their world than they should, and besides, they seemed to have some kind of sonar. Whatever it was, it made their voices sound funny.

She bowed, trying to breathe through her mouth, trying not to think of fish, or of these long, sinuous creatures moving underwater like humanoid versions of the dolphins she'd seen on holos from Earth.

She was doing what she had to do to allow her people to survive. She'd chosen the most likely path to bring them to safety and to a future.

"We are honored to be here," she said, her voice sounding raspy and her throat feeling too dry in the thin atmosphere. "May the uniting of our peoples and technology be propitious to both."

Why did this feel like defeat?

The first hint that something had gone seriously wrong had been that no one who was out that "night" had returned.

It wasn't night as such, but conventionally held as the night hours. There was no real night on Gloriana, aka Ross 128 b. It was tidally locked, with a night and a day side. The day side had permanent cloud cover, which kept it cooler than it would otherwise have been, and allowed the poles to be habitable, but still there was no night as such.

The colony, established over a hundred years ago, had decided to keep "night" and "day" as roughly the same cycles they'd kept on Earth and on the slow ship out, the better to match human circadian rhythms. Though to be fair, people worked both night and day cycles. Plants grew around the cycle, and why waste time that could be spent tending the fields? Night workers just kept a different cycle than day workers.

Only on this particular day, the night workers hadn't returned. It was late in the morning and not one of them had made it back.

Martha, newly elected as colony leader, was informed of this as she entered her office. Mike, her husband, chosen according to their genetic profiles, was a dozen years older than her, and had slid seamlessly into the role of supplies manager for the colony. He'd taken the fragmentary records of the previous administration and was as informed of what they had and what they could do with it as anyone had ever been.

Their relationship was not a warm one. Both Martha and Mike had grown to adulthood under an administration that believed the rules which had held the colony together and successful for a century were no longer needed. Both of them had experienced the failures in technology and knowledge that came with people no longer being forced to learn to maintain and equip their world. They'd chosen therefore to take the path of their grandparents and be matched according to DNA. Martha, in her early thirties, was the mother of eight of their genetic children and two womb-children, the name given to the embryos brought, frozen, in the long voyage to Gloriana.

Having done her duty for the colony, having realized that the rule of Caiden Lester, the previous ruler, had brought them to some perilous places, both genetically and in knowledge, she'd run for office under the "restore and work" platform.

And while her relationship with Mike might not be warm, or passionate in the sense of those couples you tripped upon in the odd corners of the colony, locked in embrace and looking far less embarrassed than you were to see them, they had learned to be friends.

Their passion for the colony and ensuring its survival had guaranteed that, as had their preference for a quiet, ordered life. They rarely disagreed on how to raise their children, or on things like the necessity to resume hybridization experiments for Gloriana and Earth life-forms.

He was still fit, in his early fifties, and the hair going gray at his temples seemed to enhance his attractiveness. He'd gotten up earlier than she did, as usual, and was waiting in her office with a cup of coffee and the bad news.

"What?" she asked. "None of them?"

There were no dangerous animals on Gloriana. In fact, there had been nothing but vegetation, insects, and some fish and crustaceans when humans landed. That Earth vegetation and animals could thrive on Gloriana was an unexpected boon, though what thrived and what died was strange. For instance, why had goats survived but not cows? Why chickens and not rabbits? But there was no larger indigenous life-form than a kind of palm-sized spider.

That said, Gloriana had seismic activity, and there had been work parties that had failed to return in the past, having run into lava flows or other things that damaged their equipment. The normal protocol was to send search parties for anyone who failed to make it home safe in the morning. But, "All of them?"

Mike nodded. "I've sent parties out to look for them." He paused a moment. "I had them wear radiation-protection suits."

"What?"

"All our sensors indicate that the radiation outside is very high." He hesitated again. "I think the sun had a quiet flare during the night."

Martha arrested in the act of taking her cup of coffee to her lips. It wasn't real coffee. Her grandparents said it wasn't real coffee. She remembered their complaints. But the plant, grown on Gloriana, had the same caffeine content and Martha liked the flavor. It associated in her mind with waking up. Now she took a deep breath of its caramel odor, then set the cup down. Her hand would not tremble. She would not allow it.

"A flare," she said.

Back on Earth, when the colonization of Gloriana had been planned over two hundred years ago, they had known that Ross 128 was likely to flare. To be honest, all stars were likely to flare, including Earth's own Sol. But Ross 128 had been quiet a long time, and there seemed to be little risk of it. It might flare again in a thousand years or ten thousand, but on the way there the colony would have plenty of time to establish its systems of defense, its magnetic shields, or simply to move on, further, to another and more hospitable planet.

The idea that a flare would strike within just about a century of the colony's founding seemed... unlikely.

And Martha found to her horror that she was not absolutely sure of what it would do precisely. She knew about the Carrington Event from history, and that it would have destroyed any non-hardened electronics back on Earth, but she wasn't sure what it meant for humans. "Would it have killed our people?" she asked.

Mike shook his head. For the first time he looked old, as though an immense tiredness had settled on him.

They'd done a lot in their one-year term. But he'd never looked this tired.

"No, they'll be alive," he said. "But all their equipment will be dead. For that matter, we only have buggies and tractors and radios still operating because we—because you—implemented that rule about keeping all the not-in-operation vehicles in underground garages, like our grandparents did."

She shook her head, "We." He'd told her long ago that he didn't want the burdens of public leadership, of being the one in the spotlight. But they'd always been a team. "Not that we even thought about that," she said. "I mean I thought maybe some kind of storm, or..."

"I thought of solar flares," Mike said. "Unlikely though they sounded, I knew they were possible. On the other hand, I'll confess I didn't expect one this soon. I'd...I'd intended to restore the practice of having a number of people on the mothership, where they could observe solar activity and...and the rest of the system, beyond the cloud cover, but we never—it's expensive, Martha, and I couldn't justify it. We'd have to send people up every couple of years, and we'd have had to supply them, too. Not to mention no one wanted to go and live confined in the mothership."

"I know," she said. "Caiden left a mark. His leadership wasn't just leadership, but the result of the culture. And the culture itself was dying to break out. People were held to roles and rote too long. Maybe it's natural they wanted to explode out of them and follow their bliss."

Her predecessor, Caiden Lester, had held onto a policy of what he called "making the people happy." He'd thought the colony, three generations in, was established enough, and the dangers of the world were known well enough that they could relax somewhat, and just "make people happy" by letting them pick their own professions, their own interests, their own passions. "Follow your bliss" had been one of Caiden Lester's catch phrases. He thought that now that the population in the colony had reached a thousand people, they could afford to stop interfering in personal choice in their occupations, personal choice in their mate selections, and personal choice in their free time.

While Martha, having grown up in a colony where it seemed

to her that all her movements were carefully watched and scripted, agreed that there should be more freedom, she had fought against Lester's ideas that no vigilance was needed: no mandatory observation of the environment around Gloriana in case of surprises, no mandatory genetic mapping pre-marriage, no extraordinary vigilance. With a thousand individuals in the world and ten thousand embryos still awaiting thawing and carrying to term in the mothership, the colony was now safe enough, secure enough. It needed no special hardship to keep going, Lester said. And Martha had wondered if a hundred years of stability really meant anything.

Eventually Lester's aversion to other types of planning had led to people electing Martha, who thought that crops must be planned, and that it didn't hurt young people to do a term of service in the farms, or to learn to fly shuttles that might never be needed except for the decennial ceremonial flight and the bringing down of additional embryos.

Martha knew, from reading over supply lists, that when it came to essential food supplies, Lester had brought the colony to the brink a few times. But this—

"There has been no observation?" she asked. "None whatsoever?"

Mike shook his head. He took a deep breath. "We might have to figure it out now. We might very well have to have everyone return to the mothership, and I don't know if we have room for a thousand people. But we might have to find out. Even if all the night shift died, that's two hundred individuals. The outlying farms . . . many are still underground. They should have survived, at least if they were home."

"Why?" Martha asked. "Because our equipment was destroyed?"

"No." Mike sighed. "Because a flare of the magnitude I suspect we had would have altered the atmospheric balance and likely killed all plants and animals. And made it impossible to farm this land for the next several hundreds of years."

"But it can't have," Martha said. "We know in the past these flares were much more frequent, and yet we found native plants and animals in residence."

"Yeah, but we don't know how. It's possible that there is something to local life that allows it to survive these, just like Earth life can survive volcano eruptions and other natural catastrophes. We might just have lost Earth plant life."

✧   ✧   ✧

They hadn't just lost Earth plant life.

Most tragically they'd lost all of the night shift. The colony was shocked at the loss of one fifth of its population. Everyone had lost a mother, father, sister, brother, husband or child. Some people were left alone. But worse, even as the memorial services were carried out, the evidence accumulated which made them realize that nearly all plant life had disappeared from Gloriana and the irradiated world might not recover for decades.

"Maybe centuries," Angelo said.

Lucy's eyes went very wide in response, and her mouth formed an "o." She sat in the communal refectory across from Angelo. They were both in their early twenties and it was too much to say they had been courting. Before the disaster, they had spent some time together in the sort of parties the colony threw, but they'd barely known each other.

A hundred years in, the colony had started spreading out. Families had built individual lodgings mostly dug deep into the earth, which now must be their salvation, as anyone who hadn't had a thick coat of earth between them and the surface had died.

Only a core remained, keeping up the early planned farms.

To be fair, the private farms outproduced the others twenty to one.

To also be fair, it had been more comfortable to be outflung, outspread.

Now, in the wake of the flare and with the possibility of other and possibly even worse ones—there had been a mini one, like an echo, shortly after—everyone had returned to the old quarters, the hardened, dug-in-the-earth ones created by the original colonists. They were eating food from every store they could gather, including those that had once belonged to private families, though most of it had become too irradiated to eat.

And they ate in the refectory, pretty much day-around. It was a noisy, crowded place, where you claimed a seat if you were lucky, and if not, you ate standing up by the wall. Lucy and Angelo had struck an acquaintance and usually reserved seats for each other depending on who arrived first.

"We don't have food for centuries," Lucy said.

"No," Angelo said. "We don't have a year, I hear. Old Mike told someone who told someone who told me, so it's a rumor and take it for what it's worth, but at the current rate of consumption we have maybe six months, after which comes cannibalism and death."

Lucy made a face. At twenty she'd been looking around for a husband and was ready to start a family. She'd thought she'd have five or six children, ten if she were lucky, and maybe two or three womb-children, and live in an outlying farm—a happy farm wife. And frankly, even after the cataclysm, she thought the old people, the ones who knew more, would find a way to make it happen. And perhaps Angelo, dark-eyed, brooding, muscular Angelo, would make a good husband.

Cannibalism and death just weren't in her plans. "Like that?" she asked. "No one will do anything?"

Angelo lowered his voice and leaned in. "We're doing something. I'm flying a shuttle to the mothership tonight."

"We're going to go back to the ship?" she asked.

"Can't. Not enough room. There were two hundred original colonists. We can't cram eight hundred people into it, much less supply them with the means to survive. The reactor probably could handle it, but there's not enough room. The recycling of oxygen and water for that many people might be too much. And we don't have the ability or tech to create more deep-sleep berths."

"Then why?"

"Because there is another world in this system. Ross 128 c."

"I thought that was an ice-ball. At least the instruction films—"

"It is."

"Then . . ."

"When the expedition was planned, it was thought that there were liquid oceans and life under a thick crust of ice. The thick crust of ice probably protected life in the waters from the solar flare."

"Uh, is there life under the ice?"

"We don't know. We're about to find out."

"Can you fly the shuttle?" Lucy asked.

"Technically at least," he said. "I've flown it in the simulator. Caiden Lester let it fall in disrepair. But Mike says there is enough fuel and it's good enough to get us to the mothership. And I have the best simulator record on docking."

Lucy wasn't sure where it came from. She just knew she wanted this to work. She wanted Angelo back. She wanted some way they could still have the life she'd dreamed.

She kissed him, fleetingly, on the face. "Come back to me

alive," she said. She then got up and fled before she could see his reaction.

Mike stood, dumbfounded, not sure what to tell Martha. Eventually she had to find out.

"Definitely igloos, the report says," Mike said. "We sent up the shuttle and a couple of the people who've been studying—at least theoretically—how to handle the observation telescopes. And they say the structures that cover most of the northern hemisphere are definitely artificial and they look like Earth igloos. But they could be aliens."

Martha looked tired. She gave him a faded smile. "Well, then. Do we have any choice but to meet these aliens?"

"I suspect not. We can refuel from the mothership and we estimate that a trip to Ross 128 c will take maybe ten days. It's doable in the shuttles."

"We'd better take volunteers," she said. "If they die..."

"Oh, we're going to have deaths on our conscience one way or the other. But I agree. We'll take volunteers both for pilots and for the party to meet the aliens," Mike said. "Let's hope there are enough."

There were a lot of volunteers to both fly the shuttle and go meet the aliens. Lucy was happy she made the party of greeters—the Welcome Wagon, as the colony had taken to calling them. Though, really, shouldn't it be the other way around?—and that Angelo would be flying the shuttle.

Ross 128 c was a little world, but perhaps somehow there would be room for an aquatic farm, for...herding fish or something. She didn't want to think too hard about it. She'd grown up in a farm with goats, but the goats all had to be slaughtered when the vegetation died. Now...

She'd almost gone insane the ten days of the trip. Now they were landing and she had to admit it didn't look promising. Ross 128 c was a tiny world, green-blue due to algae trapped in the ice.

They landed close enough to the structures to see they were in fact igloos. And then the natives came out.

"From Earth," Martha told Mike. "But I'm not sure I'd call them human. They were headed further on but their recycling systems started malfunctioning. Insufficient air. Something to do with the electrical system and their reactor. They didn't have the ability to repair anything and could only limp to the nearest promising world. They used their shuttles for landing, one last time. Their useless ship remains in orbit. Many of the embryos were lost. They could only bring down the essentials, and the essentials allowed the landing party to barely modify themselves for the environment, but genuinely genetically modify their descendants and the frozen embryos.

"They are amphibian. Well, not quite. Mammals, but with the possibility of storing enough air that they can maintain under-sea... cattle farms, I suppose we must call them. Their complex system of transparent igloos over other tunnels did indeed save their crops from the flare, though they suffered some losses. They offered us a trade. We can come and join them. They can modify us enough that we can survive and modify any newborns or embryos enough that... well... we can find shelter there. If we're willing to become like them."

Her reluctance must have shown in her voice, and it showed even more when she played the films the Welcome Wagon had taken to Mike. Mike's reluctance was also evident. His lip curled. He sighed. "Is it even possible to make us like that?"

"No. Not really. Not fully. But it's possible to take us halfway there, as their landing party was. Mostly bio-modified viruses that will alter our genome to survive on thinner air, to endure the cold, to be able to oxygenate our blood to the point we'll only need a breath every hour or so. The natives can go six hours without breathing." She put her head in her hands. "I don't know what to do," she said. "They will welcome us, and our embryos, as many as we're willing to trade, because they need the genetic diversity. Only fifty founding couples survived on their side, and the gen mod narrowed their gene base even more. In return we can do anything we want with their marooned ship. It's possible we can fix it. Part of their tragedy is that their main technicians were killed in the original crisis, and they don't even know what it was. Most of the survivors were people awakened from deep sleep by the emergency systems. We were a planned colony and how-ever bad Caiden Lester's administration was, we still have techs."

"Caiden wants to stay, by the way," Mike said. "He opposes all attempts at leaving the colony. He thinks the vegetation will come back soon. He thinks we're tyrants, trying to circumvent his choice."

"*Can* the vegetation come back soon?"

"Not a chance, though we might be able to provide him, and anyone who wants to remain, with the hydroponics facilities, and the seeds. Maybe one hundred people can stay behind."

"It will have to be voluntary," Martha said.

"All of it will have to be voluntary. Do you want to have any of it on your conscience?"

She laughed, but there was no joy in it. "Not even my own fate," she said.

"There's room for four hundred people in the two ships," Lucy said. "And we're surrendering half of the embryos to the people in Ross 128 c. They call it Diana."

"Are you staying?" Angelo asked, with some anxiety. "Only, I've requested to go on the mothership. Our original mothership, that is, the *Eos*. It's going further on, you know. I don't know if I'll be chosen or exactly where we're going yet. People have mentioned Kepler, I think, but I'm not one of the eggheads, so I can't tell you. The techs are fairly sure they can get us to a world, maybe better than Gloriana and close enough that, using long sleep, you and I—I mean, I'd still be young enough to start a life and a family."

Lucy smiled. "Yes, you and I could do that, Angelo. If we're keeping the domestic animal embryos . . . maybe we can still have a farm somewhere. I've requested the *Eos*, too. A lot of stick-in-the-muds want to go in the Diana colony ship and back to Earth. They called it the boomerang."

"Well, if they don't accept me to the *Eos* then maybe Earth will be okay," he said. "I hear they have enough room for farms there."

"For all of us?"

"Probably, but Lucy, if we are going to stay on the same ship when they choose, perhaps we should get married? I hear they give preference to married couples."

"Uh . . . is that a proposal?"

He leaned in close in the crowded refectory. People were talking very loudly all around, and in a corner Caiden Lester was holding forth on how a hundred of them could hold this world, and make a stand for another generation or three, and then they could recolonize. Ross 128 was a quiet star, as quiet as the sun. Another catastrophic flare was unlikely in the near future. They needed, of course, to pick those with the cleanest genetics among the colonists. One hundred people could form a viable colony, particularly if supplemented with a few frozen embryos, but they must have clean genetics. And breeding would have to be carefully controlled. Lucy spared a look at the man's haggard face, as he expounded the exact opposite of what his administration had stood for, and then leaned in closer to Angelo, in time to hear him ask, "Lucy, will you marry me?"

"Of course. Let's register it."

And maybe there wouldn't be enough time or resources for a farm in their future. Maybe like the landing-party generation in Gloriana they would have to sacrifice their lives in regimented, almost military discipline. Maybe she would have to spend most of her life pregnant.

But somewhere they would go on, and she would have Angelo. And maybe their children or grandchildren would have their own farms and find a world more hospitable than Gloriana had proven to be.

Mike waited. He stood by his wife's desk, looking at her with every air of expectancy.

He'd brought her a cup of coffee. Martha sipped the coffee as much to wait out the need to speak as to savor its rich caramel flavor. She wasn't sure there would be any equivalent where she was going. And she didn't want to go. She really didn't want to go. Given her choice she would go to Earth, or on the *Eos*. But she was too old to establish a new colony. Even with the minimal aging while in deep sleep, she might not be able to have more children. And Mike—

Beyond all that, it behooved her to provide a good example to her subordinates. Even with two ships, there was only room for four hundred people, give or take. Even with Lester's group subtracted, they'd need three hundred people to stay behind. Three hundred one if they counted the new Martin baby in the population, though he and all the young children would probably go in the *Eos*, in deep sleep.

"Patrick and Peter and James have signed up to the request list for the *Eos*," Mike said. "And I allowed Mary and Jane to make the choice as well." Patrick, Peter and James were their sons over fifteen and Mary and Jane were their ten-year-old twins.

Martha felt a wrench at her stomach. She was sending five of her children to places unknown. It was logical, even loving, to give them a chance at a new world. It was risky also.

"I suggest we send Michael, Miles, and Janet with them, as well as the babies."

Michael was five, Miles three, Janet two, and the babies, her womb-children, were one year old. The wrenching increased. "I don't want—" she started.

"If we have them modified to stay in Diana," Mike said, "they'll never fully fit in. It will be difficult for them to procure mates or to start a new life. They'll be like us, creatures caught in the middle, but for their entire lives. It will inform their ideas of themselves, everything they are. We can send them back to Earth, of course. They should be able to find a place there, but it is also denying the entire effort our ancestors made to establish a new beachhead for humanity. And we don't know what the Earth is like after two hundred years. Or how they'll be treated. They might become curious specimens to be studied and observed their entire lives."

"There is no good option," Martha said.

"No. We can have more children," Mike said. He put his hand across to her.

She looked up at him, and was both comforted and shattered to see tears in his eyes. "But you don't understand," she said. "There is only one place I can go. I have to go to Diana. Someone has to set the example, and by virtue of being the elected leader when this happened, it falls to me. The captain goes down with the ship and all that."

"I know."

"And you don't mind?" she asked. They hadn't married for love. Their married life had been harmonious rather than passionate. He'd been a reliable ally and a steady worker for her goals. But all of a sudden, inexplicably, in the middle of the destruction of the whole world, she needed to know for sure that he would be with her. They might have to cast their children adrift into the unknown but she didn't know if she could survive without him. She looked at his green-blue eyes and realized he was looking at her just as intently.

"Wither thou goest, I will go, even beneath the ice."

✧   ✧   ✧

Martha wanted to laugh and cry at once. The facilities were more technologically developed than she expected under those igloos. They should have known, of course. After all, creatures who can perform gen mods aren't exactly primitives, no matter how many fish skins they wear, nor how many ways they choose to decorate their skins with fish bones and tattoos.

The accidental, ill-begotten colony was about where the carefully planned Gloriana had been. They were at the point that there could be individual farms and some autonomy.

They retained, from the two ships, enough tech to keep a wary eye in the skies in case there should be a flare so massive it would affect them.

But now the five hundred Gloriana embryos had been delivered and the shuttles returned to the mothership.

They were here for good. Their boats had been burned. There was no going back. They could never leave this shore.

She held Mike's hand as she was put under, to start the modifications that would make her a native of Diana.

It all depends on what eyes you use to look at the world, Martha thought. She looked out the window of the igloo that was both greenhouse and home. All the plants gave the house a very high oxygen content. And the plants grew at levels that her two-year-old, Triton, couldn't reach.

She heard the splashing down from the lower level as Mike came in from tending to their fish.

He brought back a crab. It wasn't really a crab, not unless crabs were the size of dinner plates and pale violet. But it tasted much like Earth crabs. She straightened from where she had been singing Triton to sleep, and welcomed her husband with open arms.

If their movements were weird now, or if they smelled like fish, she couldn't detect it. He was still the most handsome man in the universe, and Triton was as beautiful and loved as her other children had been.

They had scattered their children to the universe and she prayed, as she did every night, that they'd been granted safe landing.

But looking around her home, filled with plants that glowed orange gold in the ice-filtered sunlight, she found nothing missing.

They would go on, as colonists had had to since the first amphibian crawled from the deeps onto dry ground on Earth millions of years ago. They would go on as human colonists had managed to throughout the centuries.

Sooner or later Earth would find faster ways of traveling—warps or gates or something—and they'd come and unite all their colonies into some sort of human commonwealth.

Martha would wager that when that time came the inhabitants of Diana would be far from the strangest, accidental or planned.

She took the crab from Mike's hands and kissed him passionately, smoothing back his soaked hair.

Perhaps Earth would come much sooner than expected. Perhaps they'd even understand her attempt at communicating they hadn't all died, and that they might be found elsewhere in the system. Or perhaps she had just sought comfort from history and from the tale of the lost colony of Roanoke.

And yet, if Lester's remnants of their people didn't make it, perhaps someone would correctly interpret the word *Croatoan*—the sole remains of the lost colony of Roanoke—which she'd ordered carved deep in the rock of what had been Gloriana as they left.

A sign to those who came after.

Bridging

William Ledbetter

William Ledbetter is a Nebula Award winning author with more than sixty speculative fiction stories and nonfiction articles published in markets such as *Analog*, *Fantasy & Science Fiction*, *Asimov's*, Baen.com, the SFWA blog, and *Ad Astra* magazine. He's been a space and technology geek since childhood and spent most of his non-writing career in the aerospace and defense industry. He administers the Jim Baen Memorial Short Story Award contest for Baen Books and the National Space Society, is a member of SFWA, the National Space Society of North Texas, a Launch Pad Astronomy workshop graduate, and is the Science Track coordinator for the FenCon convention. His science fiction thriller novel *Level Five*, published by Audible Originals, is available in audio format at audible.com. He lives near Dallas with his understanding wife, a needy dog, and two spoiled cats. Learn more at www.williamledbetter.com.

Sigvaldi and its trailing moonlet, Astrid, were already high in the dark sky, but still too far away for me to see the long thread of the bridge. They were waning and didn't provide much light, so I stepped from the shadows cast by the launch gantry and squinted into the growing dawn. My pulse quickened and the air crackled as a distant, growling roar announced the fødselsvind's approach.

Ghostly on the horizon, the pale, towering dust tsunami separated from the darkness. It raced toward me, occulting the horizon and even the stars above. I lowered my goggles, pulled the scarf up over my face and planted bare feet, shifting them until they felt firm, then crouched and leaned forward.

The dust swept in, gently at first, then building in strength. I waited until stinging grit bit my bare skin before sucking in a breath and holding it. The wind intensified, threatening to push me backward, but I leaned further into it.

I refused to yield.

I would not move my feet.

I would best the fødselsvind.

An unexpected gust made me twist, arms reeling and I almost fell. Then it was over as the terminus wind swept past and a new day was born.

I released my breath, sucked in another through my nose filters, then spread my arms wide and screamed defiance into the settling dust. More yells rang from the paling darkness as other geitbrors also proclaimed their victory and mastery over nature. Like the seven generations before me, I had proved worthy of another day on Støvhage.

Hearty laughter rang out behind me and a strong hand clamped my shoulder, raising a small dust cloud. "By the gods, Judel, will you goat herders ever become civilized?"

I turned to face Alvin Lund from the government's intelligence office and shrugged off his hand. Not only had I bristled at his suggestion that my family heritage was *barbaric*, but something about the man made me squirm in my skin anyway and I didn't want him touching me. Like most visitors from the coast who braved fødselsvind, Alvin wore full storm gear, including a sealed coat long enough to brush his boots.

"What do you want?" I said. "I know you didn't come to celebrate the new morning. Blow the fish stink off, perhaps?"

He smiled and shook his head, but didn't rise to the insult.

"This," he said and gestured around him, "is a good place to talk."

I started walking toward the dormitory, letting him know my opinion of his wanting to talk. "You're making a mistake. I'm not a spy."

His coat made a rustling, hissing sound as he scuttled up beside me. "We're not asking for much. Just tell us what you see."

"I don't spy on my friends," I said, reaching for the dormitory door handle.

He grabbed my arm, then slipped between me and the door.

"That's just the point, Judel. These people left us down here to die and ignored our struggle to survive for a full century. They are not our friends!"

"Their ancestors did that to our ancestors," I said. "What good does that grudge do anyone now?"

"They're using you, Judel. Do you think it's coincidence that your contact is a beautiful woman? Do you think for a second that Sofie isn't just flirting with you and stringing you along? These skogsrå are good at seduction. They know how to manipulate men."

"Skogsrå? So now they're magic fairies? Look, if you think I'm so easily fooled by these people, then find a new chief project engineer and send him or her instead."

"We've considered that," he said in barely more than a whisper. "You're too deeply embedded in the project. They trust you."

"Which is exactly why I don't want to betray them."

Lund's gaze grew hard in the dim morning light. "Betray *them*? Is that how you feel, Judel?"

A chill crept up my spine and for the first time since the government spooks started courting me, I felt a flicker of worry. "Believe what you like. Put me in jail if you doubt my patriotism."

Lund's expression changed immediately to one of brotherly comradery and he clapped me on the shoulder again. "We don't doubt you or we wouldn't let you go, but it's easy to lose focus on what is really important sometimes. While you are up there, or any time you are working with these Sigvaldites, remind yourself that their motivation is the same as ours. To do what is best for their *own* people. If you remember that, and view everything they do through that lens, then they can't fool you."

"I launch in about two hours," I said. "I . . . thought maybe I'd check in and see if you need for me to bring anything up."

Sofie smiled into the camera and must have answered my call from home instead of the engineering office. When she shook her head, loose hair lifted into the air and then slowly settled to her shoulders like storm-driven snow when the wind suddenly dies. So strange. She lived in an environment with which I had no experience.

"You already told me you can't sneak fresh salmon up here this trip," she smirked. "So how about bringing Frank?"

This time I shook my head and laughed. "The Space Authority refuses to let him fly without a space suit and those seem to be hard to find for dogs. You'll just have to meet him when you come down to my farm."

"I'd like that," she said with a faint smile and rested her chin in her palm. "But for now, I'll have to settle for you."

That smile, and her little teasing remarks always made my pulse race, but now Lund's comments intruded and made me wonder. Was she manipulating me? My intuition said no, but would I know? Or did I like her so much I would subconsciously overlook subtle signs?

"Do you have restaurants up there?" I said, deciding to go for it.

"Of course. Your people are supposed to be the 'barbarians,' not us."

"Then maybe, for lack of that fresh salmon, you can introduce me to your local cuisine instead."

Her eyes opened wide. "Are you asking me on a date, Judel?"

I flinched. "Well...I mean if..."

"In my opinion, we have excellent food. Even pizza," she said.

"Pizza?"

"And by the way, I'm a pretty good cook if you like vat-grown meat and algae casseroles."

The face I made must have been awful, because she laughed hard. "I also make good cookies. Maybe we should stick to those. Your stomach will probably be squishy from the flight and low gravity anyway."

A pounding on my door reminded me it was time to go and we said our goodbyes. As I let myself be led to the launch-prep area, I wondered if her comment about being a good cook was some kind of signal or hint. Was she planning to invite me to her apartment? Or wanting me to ask? And if so, should I go? Damn, Lund! He had lodged doubts in my mind like a treble fishhook.

I had little to do during the pre-launch period but lie on my back, look at the vast instrument panels, and "not touch anything." Much of the originally planned gauges and switches had been replaced by Sigvaldi touch panels. All of the computers were of their design as well. I couldn't deny that they were superior, lighter, smaller, and more efficient, but it made me uncomfortable. It was hard to argue against the systems—for spaceflight and the bridge control—needing to be seamless and integrated, but we relied on them too much. If relations with those fickle Sigvaldites went belly-up again, then we could be in trouble. Of course, if that happened then the bridge project would become a moot point anyway, so it wouldn't matter.

I closed my eyes and tried to relax the knots in my stomach. Radio chatter filled my helmet as the two pilots worked through their checklist with ground control. I wasn't afraid, not really, but my loss would set the project back by at least a year, maybe more. This entire launch was simply to get me up to Sigvaldi so I could interface with their engineers. Sofie insisted I use their immersion fields and that wasn't something they could export to us.

The pilot next to me, Anna, laid her hand on my arm and said "This is it, Judel. You ready?"

I nodded, then realized she couldn't see me in my helmet and croaked out "Yes."

A series of bumps and bangs made me flinch, then I felt and heard a roar that grew steadily deeper. Vibration made my vision blur and teeth rattle, then a load settled on my chest. Støvhage was larger than Earth, so already had a gravity that was nearly half again the evolved human normal. Those of us living on the surface had adapted. Each generation handled the added stress better, but escaping the gravity well of this monster planet required a lot of power. As a result, rockets were only built when needed and crewed flights up to Sigvaldi had been few since the colony's founding.

It was easy for the Sigvaldites to come down to the surface—they only needed a capsule that could withstand reentry—but, perversely, once they arrived they were invalids dependent on mechanical aids to even get out of bed. Only twelve had visited during the last hundred and fifty years, but due to the gee forces imparted by our chemical rockets, none were ever able to go home. I understood the problem implicitly as I struggled to stay conscious amid the building force. I—one of the strongest generations yet produced by Støvhage—could barely breathe. Hopefully, the bridge would end our dependence on rockets.

Microgravity was not kind to my stomach and even though they had trained extensively for this mission, the pilots had not actually been in space before either, so my upset triggered a "bag use" chain reaction. Luckily, they didn't hold a grudge.

On our second orbit, one of the pilots suggested I unharness and look out the tiny window. We were passing beneath Astrid, the kidney-shaped moonlet. Just beyond I could see Sigvaldi's cratered surface, brighter and clearer than I'd ever seen it from the ground. Amazing as those sights were, I was transfixed by the long glittering string of the bridge.

It was really more of a sky hook than a bridge or a space elevator, but "bridge" had been used by the media and politicians, so the term stuck. The structure was only anchored on the Sigvaldi end and trailed the moon like the leash dragged behind by a runaway dog. Its loose end flew through Støvhage's sky, skimming a mile or more above the surface and only getting close enough to access when it crossed the high plateau. That is where my work started. I'd designed the maglev system to accelerate the carriages up to a speed where they could catch and grab the end of the leash as it passed. It had been the most challenging and rewarding work of my life, yet paled in comparison to what Sofie and her team from Sigvaldi had built.

Our ship drew closer, but as interesting details began to emerge—things like moving robots and construction workers—my crewmates insisted I return to my seat for docking. We arrived at maintenance hub four, which was little more than a blister on the spine of the bridge. It contained emergency medical supplies and feed lines for fuel and volatiles, but nothing like a crew quarters. Not that it mattered. Both pilots were forbidden to leave the capsule for the entire week I'd be on Sigvaldi.

"Don't forget to come back," the flight commander said. "I'd hate to stay locked up with Anna all this time for no good reason. I mean she snores and chews with her mouth open. Very crude."

Her comment had been stated as a joke, but I knew she was quite serious. I wondered what Lund and his spooks had told them. That I was a defection risk? That I had been bewitched by skogsrå?

"Keep the engine running. I'll be back soon."

The maintenance hub was supposedly pressurized, but as a precaution I sealed my helmet before opening the docking hatch. I pushed my duffle bag ahead of me, then squeezed my bulk through as they closed it behind me. I clipped my bag to my lower back, below the environmental unit, then pulled myself along the bridge spine in a series of awkward lurches until I found the transfer hatch. The clumsy jerking around in microgravity made me dizzy and sent my stomach into queasy somersaults again. I had to gain control. I didn't want my first face-to-face introduction to the Sigvaldites—and most especially Sofie—to be amid a vomit cloud.

Lund's voice hissed in my helmet speaker. "You doing okay, Judel?"

They had all the data feeds from my suit, so they knew exactly how I was doing. The bastards even spied on their spies.

"Couldn't be better," I snapped. "This null-gravity stuff is a breeze."

He laughed. "Good. Our boards show your ride is only a couple minutes away, so hang on just a little longer."

I closed my eyes, swallowed hard, and pleaded in desperation with my stomach to not fill my helmet with puke.

"We're all counting on you, Judel. Make us proud and don't let them push you around."

"Do my best," I grunted.

Vibrations, strong enough to feel through my pressure-suit gloves, announced the carriage's approach. I wondered if the shaking was caused by an anticipated interplay between carriage and bridge, or if it was an unexpected oscillation that could be a real problem. Maybe the recent docking of our capsule set up a local resonance in the bridge structure? It bothered me not knowing all the design elements of the bridge itself. We on the surface were only responsible for the terminal interface.

A green light winked on, then the hatch slid open revealing a small airlock. When I stepped in, my space-suited bulk nearly filled the entire area. Since both sides were already pressurized, the other hatch opened almost immediately revealing Sofie, who floated beyond. She offered me a bright smile and held out her hand. I took it, intending a handshake, but even though her long fingers disappeared in my massive glove, I could still feel her strong grip as she tugged me into the carriage.

She helped me remove my helmet, then kissed me on the cheek. "Welcome to the bridge!"

I reached for her, intending to kiss her back, but she ducked and my forward momentum sent me somersaulting across ten meters of empty air. Once the bridge was operational, the carriages would all be configured for cargo or passengers, but aside from some maintenance equipment strapped to one wall and four acceleration couches bolted to the "floor," this carriage was empty and offered nothing for me to grab. I continued to spin, seeing Sofie's grin once every rotation as she tried to stabilize me. She finally caught my arm and my tumble slowed, though imparted some of that motion to her. We were both laughing, hard, by the time we slammed into the padded far wall.

Using cargo cleats on the wall, we worked together and pulled ourselves down toward the floor, then grabbed handholds on one of the couches and were strapped in within a couple minutes.

Only then, after she had turned partially away busy with her own preparations, did I let myself really look at this woman. Scale was hard to get from a video link, but in person I could see that she was at least several inches taller than I. She didn't wear a pressure suit—a probable violation of protocol—only a form-fitting utility layer that implied her weirdly long arms and legs were mostly hard muscle.

She gave a series of verbal commands. The carriage movement was at first barely perceptible, but acceleration increased until reaching about one and a half planetary gees, where it remained.

Surprised, I turned my head just enough to see Sofie. Her usual unfathomable half-smile had been replaced by a grimace and short, forced breaths. Before I could take any satisfaction in seeing one of the mighty Sigvaldites laid low, she gasped out, "Normal passenger . . . acceleration will . . . be a half gee. But I thought . . . you could . . . handle this."

Her smile returned, just for an instant.

The carriage eventually stopped at a terminal carved deep in the heart of Sigvaldi. The cavernous, echoing rooms were cut from bare stone and empty except for the occasional robot or human worker mounting signs or polishing surfaces. At least from this perspective, it little resembled the glittering faerie cities of rumor and legend.

My disappointment at this fact was compounded by the irritation at having ridden all this way inside the bridge, yet aside from the brief glimpse through the capsule port, I didn't really "see" it. Was this life on Sigvaldi? Seeing everything only from the inside?

Sofie took me to a room lined with large lockers. I was suddenly aware of my smell as Sofie helped me out of the pressure suit, but she didn't comment, as she hung it and my helmet in a locker. I was hesitant when I saw no lock of any kind on the locker, but she only smiled, took my hand and placed it against the outside of the closed door.

"It has a built-in palm reader and now will only open for you."

She demonstrated how it worked. I'd read about such technology, but had never seen it. I longed to examine the interior mechanisms but she instead gently propelled me out the door.

I was relieved to get out of the uncomfortable suit, but even happier to be out from under Lund's watchful eye. His surveillance of me was locked away with my suit. At least for now. I followed her down a long corridor to a train station also cut from the unadorned stone.

"Considering the size of those rooms and corridors, you must be expecting a lot of traffic from the bridge," I said.

For the first time since our initial meeting in the carriage, she looked me directly in the eyes and smiled.

I melted a little.

"It's easier to build everything large from the beginning than to go back and change it if needed later," she said.

I nodded, not entirely trusting myself to speak. Lund's words kept echoing in my head. I am not going to let this happen, I thought. She is just a woman, and an engineer, not some mythical fairy who can bewitch a man's mind. At least not mine.

The train arrived and we rode in an awkward silence for about fifteen minutes. When the doors opened again, it was on an entirely new world. Tall, brightly dressed people loped back and forth along the train platform. The walls were covered with tile mosaics, frescos and woven tapestries. I paused to stare, becoming the stone jutting out of a fast-moving stream as the people gave me curious glances, but flowed around me. I tried to absorb all the details for my report, but it was like trying to catalogue the movement in a kaleidoscope. I did note that there was no sign of military or defensive works.

After a couple of seconds, Sophie took my hand and led me to another corridor.

"We have small field offices on the surface at the main construction site and even in two places along the bridge, but the immersion equipment is in our main office and that's a bit of a hike from here."

She guided me along with gentle tugs and nudges. I tried not to think about her hand in mine. There was nothing flirtatious about her guiding me along in strange corridors and weak gravity, but during the years of talking to her, working and laughing together, I had built up hopes and expectations about this moment. Part of me worried that she would try to seduce me, but mostly I feared that she wouldn't.

We eventually entered doors that were marked simply "Engineering." A hallway led to a wide, colorful room, open and brightly lit. Though quite different from my own facility, it had all the markers of engineering offices everywhere. Various partially disassembled machines sat on tables and desks and wide screens displaying test data, while diagrams and conceptual drawings covered every wall. A beautiful, quarter-sized model of my bridge carriage sat in the center of the open area.

Conversations around the room tapered off and stopped as all eyes turned our way. An almost impossibly tall man, easily a head above the other giants, stood up behind a table and crossed the room in two long strides.

"Judel, this is Luther, our engineering director and my boss," Sofie said.

He gave a slight bow, looking down at me from lofty heights and offered his hand to shake. Like Sofie's, his hand was narrow with long fingers. "Sofie shouldn't have brought you here first. You must be exhausted from that brutal flight."

"Totally my doing," I said, cutting off Sofie's reply. "I insisted on coming to your office first. I'm quite interested in seeing this amazing Sigvaldite immersion tank that..."

I froze mid-sentence, my face feeling suddenly hot as I realized much too late that I'd used Sigvaldite, what they considered a derogatory term, in an actual conversation. The term was common usage down on the surface and I had been extremely careful during all these years, only to let it slip here and now.

"Oh, I'm so sorry. I just—"

Luther held up his hand and slowly shook his head. "I'm sure all of this, including our immersion tank, must look like magic to an *engineer* who has been forced to use the equivalent of stone knives and bear skins to build your end of the system."

I wasn't sure what the hell that meant, since there were no bears on Støvhage, but I knew it had been an insult and my embarrassment flared immediately into anger.

"And for future reference please refrain from using that kind of language in front of my people. The original colonists called this moon Kanin before we even arrived and that name is good enough for us. We call ourselves the Kaninish."

I shifted my feet for better balance and Sofie put a hand on my arm. I knew I was about to say things that might tank the whole project. I also knew that wouldn't stop me. Those early days on Støvhage had been utter hell with more than half of the original colonists dying in the first two years. The world truly was dead, its soil being closer to regolith on Earth's moon than the rich dirt on our mother world. We'd known before arriving that we'd need massive amounts of nitrogen for the crops, but those who stayed on the moon to mine stopped sending it just as they refused to send down the second wave of colonists. As our pleas and protests grew more insistent they eventually cut off radio contact. They had written us off as a failure and decided to save their resources for some future attempt. But we didn't fail.

"I don't give a damn if you find that name offensive," I said. "You cut us off, leaving us to die down there with no help or resources. Your level of betrayal made Jarl Sigvaldi's seem like a childish prank, but that is the best name we had for you."

Luther's face darkened. "Those sins belonged to our ancestors, not us, and we're trying to make amends for them now, with the bridge and our technology."

"As well you should, but you've earned the traitor's name and we will always refer to you that way. If you don't like it, then send me home and try to build your bridge without us."

Luther's mouth clamped shut in a tight, white line. He glanced at Sofie, then turned and stalked away.

Sofie's already pale face took on a whole new pallor. "Well, that could have gone better. Let me show you to your room. We'll start fresh tomorrow."

"No, I'd like to see this immersion field now."

She stared down at me for a full second, then shrugged and looked around. "Alright, but we've found a potential problem. Are you in the proper frame of mind to look at this information objectively?"

"I can separate my work from my personal feelings."

She shrugged, then led me to a large room with white walls

that sparkled with thousands of tiny glittering pinpricks and handed me a pair of goggles.

"Don't remove these while in the simulation or the lasers will blind you."

The goggles were a tight fit on my broad face, but covered my eyes well enough. Sofie flipped some switches near the door and the walls flickered faster and faster until vague, ghostly structures took shape in the room's center.

As the holograms solidified, I recognized the mag rail twin track system we'd built on Støvhage's central plateau. It looked so real! Like flying over it in an aircraft. Then the bridge appeared on the horizon. Sigvaldi's tail it had been called, with the computer-controlled stabilizers providing the barb, its upper reaches disappearing into the blue sky, like the real thing. The camera view shifted slightly to focus on and follow two carriages, one on each track, accelerating toward us.

The bridge loomed larger in the background, getting ever closer, until it converged with the carriage at the docking terminus. I held my breath as the view zoomed in close enough for me to see the mechanical latch system engage, yanking the carriages off the track and up the bridge rails toward the moon passing above. But then something unexpected happened. The carriage closest to my perspective shifted suddenly, canting at a strange angle, almost faster than my eye could follow; it broke free and spun off in a ballistic arc. The second carriage then broke free and followed its predecessor in a spinning plunge to the ground below.

"What the hell!" I whipped around to find Sofie, but she was invisible in the simulation. I almost removed my glasses, then remembered it could blind me. "Where are you?"

The simulation halted and faded, revealing Sofie standing next to the control panel. "Can you—"

I crossed the room, nearly losing my balance in the lower gravity, until I was right next to her. "What is this insanity? Why did your little cartoon show such a catastrophic failure?"

"Because in the right circumstances—"

My arms flailed and spittle sprayed; I didn't seem to be able to control myself. "We've tested that design fifty times—"

Sofie not only held her ground, but bent closer to my face and put her hands on my shoulders. "Judel! Listen to me. The numbers don't lie. In a fully loaded condition, those locks will fail."

I wanted to yank away, scream in her face and leave, but there was a pleading in her eyes. She believed what she said. But those locks and that carriage design were our last bastion of respect. The entire bridge had been their idea, their design and built by them. We had built the mag rail track needed to accelerate the carriages to translation speed, but the engineering had been theirs. The carriages were our only real contribution and the locks were my design. They had wanted magnetic locks from the beginning, but we—no I—had insisted that mechanical locks would be more robust and reliable.

She squeezed my shoulders and bent a little lower until her face was inches from my own. "Please trust me, Judel."

"But we tested them," I muttered and took a deep shuddering breath. "To three times the load requirements."

She dropped her hands and straightened to her full height. "I know. But you tested for individual stresses, not all twelve factors at the same time, because there is no way to do that until the actual system is built. We also suspect the purity level of the carbon alloy in your electronic models is higher than you're capable of manufacturing, so we set it at a more realistic level in our simulation."

"But we—"

"And before you get defensive, we can't produce that level of purity either. It would only be possible with nano-assemblers and you know we haven't had any luck with that."

"Okay," I said. "Show me."

We watched the simulation four more times, with color enhancement and data tags showing exactly what happened as the stress loads piled up until the latch mechanism failed. So many thoughts crowded my head. I had doubts, but I also didn't believe that Sofie was frivolous and shallow enough to waste so much effort creating an elaborate ruse. Especially not just to get her way on magnetic locks or for a little one-upmanship. Once the lasers were shut off I removed my glasses and rubbed my eyes.

"You must be exhausted," Sofie said. "Are you ready to go to your room?"

I nodded and let her lead me through the now dark and empty engineering center and down the wide corridor to the dormitory and my room. We both hesitated awkwardly in the open door, but when I asked her to come in, she smiled and gave me a little shove. "Get some rest."

And I did just that, still in my clothes, and the last thought in my tired mind was of Sofie's hands on my shoulders, only this time, instead of yelling, she kissed me.

The next morning I awoke stiff and unsure of the time. I took a shower that required me to squeeze through a rubbery, sphincter-like membrane into a tight little closet. The water spray came from three sides and, due to the weak gravity, filled the space with a weird mist made of large droplets that clung to me and had to be squeegeed off. I put on fresh clothes and settled down to reexamine the latch design. Each time I looked at the numbers it felt like touching an open wound.

So I left my room, curious as to how far I'd get before their security wonks collected me. I wandered aimlessly for more than an hour, but the only thing I noted that could be considered useful intelligence was that most of the plazas and corridors I found were almost empty. Of course other than my dealings with Sofie, I had no clue about the schedules these people kept, so I could be between shift changes or maybe it was a holiday. I couldn't escape the feeling that the infrastructure had been designed for a much larger population.

The expected hand on my shoulder finally came, but when I turned, it was a winded Sofie.

"Good morning," I said. "Rough night?"

"Why are you out wandering around? Are you trying to get me in trouble? I'm supposed to be your escort."

"I'm a tourist. Was I supposed to stay in my room?" I said, feigning innocence.

She sighed and shrugged. "I don't know. I'm winging this too. Look, they expect us at the engineering center in an hour, but there are no actual design meetings scheduled until the afternoon. Do you feel up to a little field trip to the surface? Since you are a tourist and all, I thought you might like to see the bridge from the outside."

"Absolutely!"

She led me back to the locker room where I'd left my suit the day before. We helped each other dress and cross-checked our equipment after topping off our gasses, then walked up a long, sloping corridor that ended in a cluster of airlocks. As we waited for the air to evacuate, we checked our communications links.

"As soon as we're outside, you need to clip your short tether to the guide line," she said. "It's about one tenth of Støvhage's gravity. It's harder to reach escape velocity than you might think, but if you get a good enough launch it will still take hours for you to come down."

She took my hand before the hatch opened and held onto me through the whole tethering process. It felt a little silly, like a child being mothered, but I also knew that she'd been born in this gravity and had far more vacuum experience than I, so was grateful for the help.

We walked perhaps half a kilometer, then stopped and turned back to look at the bridge.

"I knew you'd want to see it from the outside, too," Sofie said.

The terminal building we had exited was huge, perhaps three stories high, covering several acres and that was only the end where the bridge was anchored. Aside from my brief peek through the capsule window, I had only seen the bridge from Støvhage's surface, moving past at over a thousand kph, which gave it the impression of some vast, bizarre aircraft. From this perspective it was a tower rising impossibly high into the blackness, getting ever smaller until it vanished from view.

The inflexible neck on my suit forced me to lean backward in order to look up, but upon seeing a slowly rotating Støvhage nearly filling the sky overhead I totally forgot about the bridge. Painfully white clouds covered much of the surface, but I could see green bands separating the blue of lakes and oceans from the yellow-brown of the still dead lands. The fragile proof of our generations-long struggle. It was beautiful, powerful, and compelling.

I gasped and muttered an old litany from the days when—due in no small part to these traitors on Sigvaldi—humanity's survival on the surface had been doubtful.

Sofie had said nothing during this time and when I turned to see why I found her facing away from me, staring at what looked like a smooth, nearly cylindrical mountain. Various antenna and other equipment protruded from the top and I realized it was an enormous building, covered by a thick layer of regolith.

I touched her shoulder. When she turned to face me, the lights inside her helmet glistened from tear tracks on her cheeks.

"Sofie?"

"We have to stop this, Judel."

"What?"

She pointed to her helmet and then held up a gloved hand with two fingers, then three, then one.

I changed the channel on my suit radio to 231.

"I'm going to do something stupid," she said. "Please trust me. This needs to be done." She unhooked our tethers from the guideline, then clipped them together and motioned for me to follow.

We left the marked path and struck out directly for the strange buried building. Sofie never said we needed to hurry, but I felt an urgency in her actions so I followed without comment. After about ten minutes and little talking, we arrived at a tube that disappeared into the "hillside" which Sofie entered without slowing. Small, dust-covered lights cast dim illumination on the corrugated interior which went about forty feet before ending at a very strange airlock.

The hatch was flush with its surrounding wall, revealed only by a seam not much wider than a human hair. Sofie punched a code into a lighted touch panel beside it which was totally smooth and almost completely blended into the wall. This building had obviously been here for many years, yet this was some of the most advanced workmanship I'd ever seen. Their techniques and capabilities were even more advanced than we'd thought.

The hatch sank slightly into the wall, then slid into a recess, allowing us to enter. My heart sank as I looked around the inside of the airlock. If they were able to employ such exquisite workmanship on something as utilitarian as an airlock then we were doomed. The weaponry at their disposal must be beyond our imaginations.

We passed through a powerful air curtain and vacuum system, then entered into a locker room and helped each other out of our suits. They were almost clean and I wondered why they would use such a wonderful system in this building, but not the main terminal building.

"You said we're doing something stupid. Why? What is this building?"

Sofie turned and smiled at me. "You haven't figured it out yet, Judel? We're inside the ship. This is the *Amundsen*."

I steadied myself against the wall as vertigo made the room

spin. Impossible. It couldn't be true. We'd been led to believe the ship had long ago been scavenged for its systems and materials. Yet it all suddenly made sense. The signage was Norwegian, but contained words I didn't recognize. The workmanship. The technology.

I saw Sofie in a new light. She knew I was a spy. She had to know. Yet she was showing me a secret greater than all others. "Why?"

"Wait and maybe it will make better sense when I show you the rest."

As I followed her through the ship, I saw that it had indeed been at least partially dismantled. Cavernous, echoing chambers where whole decks had been removed, leaving only wiring and pipe stubs. Hatches had been removed and I could see into large open areas and stripped cabins. We went down ladder after ladder, going far deeper into the ship and making it obvious that only a small portion made up the buried part I could see from the surface.

When we entered another echoing chamber that curved gently away in two directions—obviously part of the rotating habitat—I finally got a sense of scale. The ship had to be a kilometer in diameter and at least half of the floor space I could see was covered by sleeper units. As we threaded our way between them I noticed that some were lit up. They were all occupied . . . by babies.

Sofie led me to the only free-standing structure amid the cemetery of electric coffins and opened the door. Two surprised men stood in a room filled with humming equipment. They immediately backed away to the far wall and started talking animatedly.

"We don't have much time, so please listen and try to believe everything I tell you. We know you're a spy. I also know you are a good person and I think if you have enough information you'll make the right decision."

Despite the earnest expression on her face, I couldn't help but hear Lund's warning. *Remind yourself that their motivation is the same as ours. To do what is best for their own people.*

"Your government is building twelve large lift vehicles. We estimate each one could carry a hundred or more soldiers. I've seen the pictures, but couldn't get copies to show you. You'll just have to trust me that they exist."

A sinking feeling settled over me. If that was a lie, then it was one that rang true. It sounded too much like Lund and his people. They needed me and my intelligence report, yet they would never tell me why it was important.

"It's hard to be certain without people on the ground, but by the level of activity around the sites, we estimate they are at least a year from completion. That's why we're trying so hard to finish the bridge first. We're hoping that our people will fare better if integration is slow and natural instead of sudden and by force."

My thoughts were still on Lund and his subterfuge, so her comment took a second to sink in. "Wait. Integration?"

"Yes. It's inevitable. That's why I brought you here."

She led me to a large console, covered with data-filled display screens. "I'm not exactly sure what our ancestors originally named this machine, but we call it the DNA Adjustor. A combination of our limited population, cosmic radiation, and the very low gravity inflicts our children with an extremely high birth defect rate for natural births. So instead of normal gestation, we put the fertilized eggs into sleeper units, where the DNA Adjuster can monitor the genetic health of the babies and correct problems early."

Growing babies inside an artificial environment made my skin crawl, but the engineer in me was still fascinated. I leaned in for a closer look and marveled at the symbols and text scrolling across the screens. History told us that our ancestors from Earth directly manipulated genetics, just as they'd been able to manipulate matter on a molecular level, but here it was directly in front of me; one of those magic machines kept working for hundreds of years.

"It sounds as if you have a working system. So why the limited population? And what is the limit?"

She was slow to respond and seemed to be struggling with the question. Then I heard clattering from behind us and looked back toward the door we'd entered. A squad of armed soldiers poured through the hatch, spread out and moved toward us.

"Our population is fixed at nine thousand," she said abruptly. "That is the most our resources and economy can support. Like everything on the *Amundsen*, this DNA Adjustor machine was triply redundant. But two of the three machines have failed in the last few years. When this last one fails—and it will eventually because we don't understand the technology—then our already small population will start to collapse."

By that point we were surrounded and an earnest, yet nervous young woman approached wearing an officer's uniform. "You're in a restricted area. Please come with us."

"I still want to meet your dog," Sofie said and kissed me on the cheek.

The rest of that night I was kept locked in my quarters. I worried about Sofie—feeling somehow responsible and powerless—so I did what I always do when I'm stressed. I worked. In the morning I sent a message to Sofie and Luther, telling them I had a possible solution to the carriage-lock problem and I hoped they would listen to me.

Four armed guards arrived and escorted me to the engineering center. They even accompanied me into the large conference room, where they fixed me with menacing glares from their posts near the door.

I put my presentation up on the wall screen, ready to discuss by the time Luther arrived flanked by two other engineers.

"Where's Sofie?" I tried to look as menacing as I could without bringing the guards down on me.

"She . . . is being detained," Luther said with the same level of disgust most reserved for discussions of bodily functions. "There will be an investigation as to why she violated security protocols. In the meantime, in order to have as minimal a schedule impact as possible, we need to get these design changes approved so that the failed mechanical latching system can be scrapped in favor of our original magnetic locking system."

"As Sofie demonstrated yesterday, the mechanical latch system does fail six percent of the time under maximum loading, which is an unacceptable risk. However, problems with the mag lock system still exist too and—"

"If you're going to suggest that we don't load the system to its design maximum, then—"

I slapped the table with enough force to not only shut Luther up and make the guards take a step forward, but to raise me several inches out of my seat. "Hear me out! Regardless of what you think of me and all grounders, please just consider this simple alternative. I propose that we do both."

Luther glowered, crossed his arms and sat back in his seat. "We brought you all the way up here. Of course we'll listen."

"Our biggest concern with magnetic locks are power requirements, keeping the pads on the carriages clean and the lack of mechanical fail-safes should the system lose power. We can divert power from the lift drive to the mag locks for two tenths of a second, during the actual momentum shift when the mechanical locks fail, providing that extra robustness, then we engage the mechanical locks and shift power back to the drive system and start the lift."

Luther stared at the wall screen with narrowed eyes and I could see the gears of his mind whirling. He might be an asshole and a manager, but according to Sofie, he was also an excellent engineer.

"This way we don't have to add a dedicated mag lock power system and we keep the mechanical lock system," I said. "We just have to add the controllers and mag pads to the bridge and carriages. A far smaller design change."

Luther's lackeys smiled and glanced at each other. He sighed and said, "We'll work up some new stress and power models for this option and let you know."

Once back home on Støvhage, the flight medical team seemed satisfied with my progress, so they handed me some thick gloppy stuff to drink and left me alone. Aside from my stomach being a jittery mess and my head throbbing like a dust rock bass line, I had survived reentry without a problem. At least until the door opened and Lund slipped in. And that probably explained why they immediately separated me from the flight crew after we landed.

"Hello, Judel."

"Hi, Alvin. What a surprise to see you here." I had to be careful. Too much vitriol and he would doubt that I'd be so cooperative. Not enough and he'd suspect I was up to something.

He raised an eyebrow. "We're obviously concerned that the mission was cut short and you were sent home due to a security violation. Care to explain?"

"You're the intelligence expert. Shouldn't you have this information already?"

Lund sat down and crossed his arms. "We don't have assets among their people. That's why your trip was so important. We're limited to what our signal processing people can glean from their electronic communications."

I shrugged, took another drink of the nasty goo and made the appropriate face. "They had confined me to certain corridors and sections. Sofie took me to unauthorized sections and we both got in trouble. She was thrown in prison and they asked me to leave."

His eyes flared with excitement before he locked the stoicism in place again. I knew he wanted to ask what I'd seen, but he held it in check. "Why did she betray her own security to show you things she shouldn't?"

"It could be the fact that she, and they, know about your twelve heavy launch vehicles and the plan to put troops on their moon."

He flinched and went pale, but said nothing.

I'd caught him by surprise. Hopefully it would rattle him enough to believe the rest of my lies.

"She wanted me to see their preparations and hopefully avoid an invasion, but their military wants us to come. And be wiped out to the last man. It would evidently be an effective deterrent against future aggression."

"And what did you see?"

"She led me out of my section via a service tunnel. Every other corridor and section we visited contained armed soldiers. I saw hundreds. She said they'd been expecting an invasion for a long time and their military service is compulsory for every young person. If she told the truth, they have about forty thousand armed and active."

I surprised him again. He blinked and then leaned forward. "Their population is that high?"

"All I know is what I saw. The corridors she took me to were shoulder-to-shoulder people. It was like a hive. Maybe what you said about them breeding like animals is true."

He gave me a sick leering grin. "And is it true? Did you finally consummate your little long-distance love affair?"

I glared at him. My disgust wasn't in the slightest bit faked.

He rolled his eyes and snorted. "So. They just let you go? Knowing what you'd seen."

"That part baffles me, too," I shrugged. "I really don't know, but I suspect if they kept me, it would've given us a good excuse to attack and they'd prefer it to appear unprovoked. Maybe Sofie knew what she was doing and forced their hand?"

He stared at me for a long time without speaking. With his mask back in place he said, "You went out on the surface the night before you left. Why?"

I considered telling him a half lie, about visiting the *Amundsen* and seeing a super weapon. Maybe the old ship's asteroid deflection beam I'd read about, but instead decided it would be better to keep its existence secret. I could see these dumbasses firing a barrage of missiles to destroy it and we'd lose that wonderful technology due to lies and stupidity.

"To see the bridgehead from the outside," I said. "That is one of the main reasons I went, remember? But if they had military facilities out there, they were well disguised. I saw nothing out of the ordinary."

"Hidden from our telescopes and probes," he said with a knowing nod then stood up. "I may be back later for more details. But I have a lot to do based on the information you provided."

"I bet," I said and smothered a smile.

The goats arrived, bells jingling and bleats filling the air as I set out the last salt block. Only a few of them rushed over to lick the salt, which was a good sign. It meant they were probably getting enough. The rest milled around my legs hoping I had something more interesting and tasty in the cart.

Frank barked and circled us all, bouncing on his front legs and tail wagging madly. Then he stopped, raised his ears and took off at full speed toward the house. I looked that way but saw nothing. In an irrational move driven by hope, I pulled the AllBox from my pocket to check for messages, but it still displayed a red X meaning that my home repeater was out of range. I sighed and looked out at my inherited family farm. My gaze fell on rolling hills spotty with tough grass and knee-high bushes. Something an Earther would call scrub or trash or ugly, but considering that this world had been devoid of all life when we arrived, I thought it was quite beautiful. Someday, this would all be grass, and covered in cattle and need fences, but not in my lifetime.

I could hear Frank barking in the distance, so I nudged the goats aside, crawled onto the cart, and started back toward the house to see why he was raising such hell. As I topped the next rise I saw him circling what could only be a Sigvaldite wearing one of those spider-like helper suits. My pulse quickened. We saw many more Sigvaldites since the bridge was completed, but if one were visiting me out here then it was either very good news or very bad.

I pushed the cart to max speed and bounced over the rocky

ground until I could skid to a stop a few yards from the contraption. The occupant had lowered their composite frame close to the ground and was petting and cooing over a quite enthusiastic Frank. I recognized her voice immediately.

I knelt next to them. "Sofie?"

When she looked up there were tears in her eyes and dust turned to mud smudged on her cheeks.

"Hi, Judel." Her voice, while familiar, also held the strained wheeze common to Sigvaldites visiting the surface.

"I didn't know you were out of jail. When did that happen? I've been watching my... I've been kind of wondering."

"Two days ago. I should have let you know, but I wanted to come down here. I thought if you knew, you might tell me not to come."

"Of course not. But this has to be hard on you."

"You said I could meet your dog after the bridge was finished. I'm here to make you keep that promise."

I knelt down and wrapped my arms around Frank's neck. "Did you hear that, Frank? You, my boy, are a very lucky dog. This pretty lady came down from space and all the way out into the dusty badlands to meet you."

Frank bounced with excitement and licked my face in reply.

"I might have had other reasons, too," she wheezed. "The last time didn't really work out as I'd hoped."

A lump formed in my throat. "Yeah, me either."

We started a slow walk back to the house and the wind picked up, surrounding us with dust devils. Sofie gasped and stopped. "Will they hurt us?"

"No. Too weak," I said with a laugh. "Like if you were to try and pick me up."

She smirked and shook her head. "Some things never change. By the way, I love the solution you came up with for the bridge carriage locks. I've really missed working this last year, so indulge me and tell me what you've been working on. Your next big super project!"

"The bridge will be my last project for a while. Upon the insistence of the Government Security Agency, more specifically Alvin Lund, I've been fired and pretty much blackballed."

"Oh no. Why?"

"Once the bridge was finished and their agents started snooping around up on Sigvaldi, they realized I had unabashedly lied about your forty thousand troops."

Her eyes opened wide. "Forty thousand troops? Did you really do that?"

I nodded.

She started laughing, then stopped and gasped for breath. I didn't know what to do and fluttered around her like a scared hen. With elaborate hand gestures she waved me away and eventually regained her breath.

"Don't get me wrong," I said. "I've missed you and I'm very, very glad you came, but it sounds like this gravity is killing you. When do you go back?"

"You missed me?" she asked with a raised eyebrow.

"Hell yes, I missed you! Talking to you every day was the best part of those difficult years we spent designing the bridge. Then just as I thought... I mean you were suddenly gone."

"It must look like I'm an invalid down here, but I can get in and out of this contraption on my own. And tend to my basic needs, so I'd like to stay. At least for a little while. If that's okay. And if it's okay with Frank."

"Of course it's okay. It will always be okay."

"Pappa?"

Her voice was muffled by the scarf around her mouth, but I thought I detected a trace of worry.

"What is it, Squeaker? Are you scared of the dark?"

"I'm a little scared," she said. "But not of the dark. Some of the kids at school said the fødselsvind picks up kids and blows them away."

I wrapped my arms around her and squeezed. "That isn't true. If the wind blew little kids away, then there would be news reports about it on the feeds all the time. Besides, you're six now and very tall even for six, so the worst that could happen is it would knock you off your feet. And I'll be here with you to make sure that doesn't happen."

"Okay. But I'm just newly six. Remember that."

I tried to stifle my laugh and made sure her goggles and scarf were tight. "I'll remember that," I said. "We can go back in the house and do this some other time if you're not ready."

"I think I'm ready. Will mamma be proud of me?"

I could see the pale wall rising in the east and could faintly hear the building roar. I shifted my bare feet, making sure of my footing and positioned my daughter in front of my legs. "Oh yes, Squeaker. She is always proud of you. She was very excited when you told her."

"Are we still going up the bridge to see her tomorrow? To celebrate?"

When Sophie got pregnant, she insisted that the baby be gestated in Støvhage's full gravity; even then it nearly killed her on multiple occasions. The birth had been so hard on her already frail constitution, that she could never come down the gravity well again. Long-distance marriages were difficult and Squeaker and I missed her terribly, but the three of us made it work. "Yes, baby." I tightened my arms around her as the wind picked up. "We'll see her tomorrow."

Her hands tensed on mine as the dust wall towered over us and the wind howled. "You and mamma built that bridge," she yelled.

"Yes. We sure did."

The Future of Intelligent Life in the Cosmos

Martin Rees

Martin Rees is a space scientist and cosmologist based in Cambridge, UK. He has the title of Astronomer Royal and has studied black holes, galaxy formation and the multiverse. He has also written extensively about science policy. His latest book is entitled *On the Future: Prospects for Humanity* (Princeton University Press).

Exobiology is a burgeoning research field. In recent years, the discovery of exoplanets has proven both transformative and morale-boosting; furthermore, implying billions of locations in our galaxy where signs of life could have emerged. There is a realistic prospect that in a decade or two we will have the capability to detect biosignatures: nonequilibrium atmospheric chemistry, vegetation, etc. Undoubtedly, the most interesting of all would be techno-signatures: electromagnetic transmissions that suggest evidence of intelligence, or manifestly artificial artifacts. So, what might we expect to detect if the Search for Extraterrestrial Intelligence (SETI) is successful?

If intelligence emerges on another Earthlike planet, we are unlikely to have any perception of its nature or motives. So, I will offer conjectures on Earth's post-human future, where we can theorize the psychology of the dominant species and speculate future technological trends. The most likely scenario will depend on three advancing technologies: Biotech—issues of genetic modification and whether humans can redesign themselves; Artificial

Intelligence (AI)—generalized machine learning, and the extent to which machines become adept at sensing and interacting with the environment; and space technology—specifically, new propulsion systems. Moreover, it matters which of these advances fastest. The order in which technologies develop could make a big difference.

First, a cosmic cameo. Let us suppose that aliens existed and were watching our planet since it formed. What would they have seen? Over most of that vast time span, Earth's appearance altered gradually: Continents drifted; ice cover waxed and waned; successive species emerged, evolved, and became extinct. Yet, in a small part of Earth's history—the last hundred centuries—patterns of vegetation altered much faster than before. This signaled the start of agriculture, followed by urbanization. The changes accelerated as human populations increased. Then, rapid changes continued to occur. Within a century, the amount of carbon dioxide in the atmosphere began to rise abnormally fast. There were anomalous radio emissions, and something else unprecedented happened: Rockets launched from the planet's surface and escaped the biosphere completely. Some rockets were propelled into orbits around Earth, while others journeyed to the Moon and surrounding planets.

Thus, humans became more conspicuous. If the hypothetical aliens continued to watch our planet, what would they witness in the next century? Will a final spasm be followed by silence? Will the planet's ecology stabilize? Could there be massive terraforming? Moreover, will an armada of rockets launched from Earth spawn a new source for life elsewhere? Think specifically about the future of space exploration: Remember that Cassini, New Horizons, and Rosetta were all last-century technology. Recall the computers and mobile phones of the 1990s and realize how much better we can do today. During the twenty-first century, the entire solar system—planets, moons, and asteroids—will be explored by flotillas of tiny robotic craft.

The next step will be the deployment of robotic fabricators in space that can build large structures. For example, giant successors to the James Webb Space Telescope (JWST) will have immense gossamer-thin mirrors assembled under zero gravity. These structures will further enhance our imaging of exoplanets as well as the cosmos. Will there be a role for humans? It cannot be denied that NASA's Curiosity, trundling across a giant Martian crater, may have missed startling discoveries that no human geologist could overlook. However, machine learning is advancing fast, as is sensor technology. In contrast, the cost gap between manned and unmanned missions remains vast. The practical case for manned spaceflight gets increasingly weaker with each advance in robots and miniaturization.

The motive for the Apollo program was superpower rivalry. In the 1960s, NASA received about four percent of the federal budget, against 0.6 percent today. If there were a revival of the Apollo spirit and a renewed urge to build on its legacy by manned projects, a permanently inhabited lunar base would be one option. (An especially propitious site is the Shackleton crater, at the lunar South Pole—twenty-one kilometers across with a rim standing at four kilometers high. Because of the special location, its rim is always in sunlight; therefore, it escapes the broad monthly temperature range experienced on almost all the Moon's surface. Moreover, there may be a large amount of ice inside its perpetually dark interior—crucial for sustaining a colony. Therefore, it would make sense to build on the side of the Moon that faces Earth. However, there is one exception: Radio astronomers suggest placing a substantial telescope on the far side of the moon to shield it from artificial emissions from the Earth.)

Since Apollo, NASA's manned program has been financially constrained and impeded by public and political pressure into being exceedingly risk-averse. The space shuttle failed twice in a total of 135 launches. Astronauts, or test pilots, would willingly accept this level of risk—less than two percent; however, the shuttle had, unwisely, been promoted as a safe vehicle for civilians. Ultimately, each failure caused a national trauma. This event was followed by a hiatus in which costly efforts were made (with very limited effect) to reduce the risk in the future.

Due to its "safety culture," NASA will confront political obstacles in achieving any grand goal within a feasible budget. Additionally, it is understood that China has the resources, the dirigiste government, and possibly the inclination to undertake an Apollo-style program. They already plan to land on the far side of the Moon; however, a more ambitious plan would involve footsteps on Mars. If China desired to assert its powerful status with a "space spectacular" to assert parity, they would need to vastly surpass what the United States achieved fifty years earlier.

China aside, the future of manned spaceflight lies with privately funded adventurers that are prepared to participate in a competitive program far riskier than Western nations could impose on publicly supported civilians. In the future, SpaceX and Blue Origin will offer orbital flights to paying customers. There are plans for week-long trips around the far side of the Moon—voyaging farther (and longer) from Earth than anyone has before. Such ventures—bringing a Silicon Valley culture into a domain long dominated by NASA and a few aerospace conglomerates—have innovated and improved rocketry faster than NASA or the ESA—the latest Falcon rocket can carry a fifty-ton payload into orbit. Thus, the future role of national agencies will be attenuated and more akin to an airport rather than an airline.

If I were an American, I would not support NASA's manned program; I would argue that inspirationally led private companies should lead all manned missions as competitive, high-risk ventures. There would still be many volunteers—some, perhaps, accepting one-way tickets—driven by similar motives as early explorers, mountaineers, and the like. Therefore, the phrase "space tourism" should be avoided. It convinces people to believe that such ventures are routine and low-risk. If that is the perception, then the inevitable accidents will be as traumatic as those of the space shuttle. These exploits must be advertised as dangerous sports or intrepid exploration.

The most crucial impediment to spaceflight, in Earth's orbit and those venturing further, stems from the intrinsic inefficiency of chemical fuel as well as the requirement to carry a fuel weight that far exceeds that of the payload. If we are dependent on chemical fuels, interplanetary travel will remain a challenge. Nuclear power could be transformative; therefore, by allowing much higher in-course speeds, it would drastically cut the transit times in the solar system. This would not only reduce the astronauts' boredom, but their exposure to damaging radiation as well.

Also, if the fuel supply could remain on the ground, that would prove more efficient than if it were carried into space. For example, a spacecraft could be propelled into orbit by a device acting as a "space elevator," then a Starshot-type laser would deploy from Earth; hence, applying force to a light-reflective sail attached to the spacecraft.

Incidentally, more efficient on-board fuel could transform

manned spaceflight from high-precision to an almost unskilled operation. Driving a car would be a difficult enterprise if one had to program the entire journey in close detail beforehand, with minimal opportunities for steering along the way. If there were an abundance of fuel for mid-course corrections (and to brake and accelerate at will), then interplanetary navigation would be a low-skill task—simpler, even, than steering a car, or ship, as the destination is always clear.

By 2100, thrill-seekers in the mold of Felix Baumgartner (the Austrian skydiver who, in 2012, broke the sound barrier in free fall from a high-altitude balloon) may have established bases independent from Earth, Mars, or maybe on asteroids. Elon Musk has expressed his desire to die on Mars—but not on impact. This is a realistic goal that may be alluring to some.

However, do not expect mass emigration from Earth. This is where I strongly disagree with Musk, who advocates rapid buildup of large-scale Martian communities. It is a dangerous delusion to think that space offers an escape from Earth's problems. We must solve these issues here. Coping with climate change may seem daunting, but it is a small plight compared to terraforming Mars. Nowhere else in the solar system offers an environment even as clement as the Arctic, or the top of Mount Everest. There is no "Planet B" for ordinary risk-averse people. Nevertheless, we should support those brave enough to venture into space, because they will have a pivotal role in leading the post-human future and determining what happens in the twenty-second century and beyond.

The pioneering explorers will be unsuited to their new habitat, sustaining a more compelling incentive to adjust themselves compared to those of us still on Earth. They will harness the powerful genetic and cybernetic technologies that will be developed in future decades. These techniques will be heavily regulated on Earth as well as on prudential and ethical grounds; however, settlers on Mars will exceed the clutches of regulators. Therefore, we should wish them luck in modifying their progeny to adapt to alien environments, as this might be the first step toward divergence into a new species. Ultimately, it will be these brave space voyagers who lead the post-human era.

Organic creatures need a planetary surface environment; yet, if post-humans make the transition to fully inorganic intelligences, they will not need an atmosphere. Rather, they may prefer zero-gravity, especially for constructing massive artifacts. So, it is in deep space—neither on Earth, nor Mars—that nonbiological "brains" may develop powers that humans cannot begin to fathom.

In fact, there are chemical and metabolic limits to the size and processing power of organic brains. We may be close to these boundaries already, but no such limits constrain electronic computers, much less quantum computers. So, by any definition of thinking, the amount and intensity that is done by organic, human-type brains will be utterly swamped by the cerebrations of AI. So, although we may be near the end of Darwinian evolution, the technological evolution of intelligent beings is just beginning. It will happen fastest away from Earth—I would not expect such rapid changes in humanity here; our survival, though, will depend on ensuring that AI on Earth remains benevolent.

Few doubt that machines will gradually enhance or surpass our distinctively human capabilities. There are disagreements about the timescale. For example, Ray Kurzweil and those who hold similar opinions claim it will be a few decades; in contrast, others envision centuries. At any rate, the timescales for technological advances are instantly compared to the timescales of Darwinian selection that led to humanity's emergence. Therefore, the outcomes of future technological evolution could surpass humans by as much as we (intellectually) surpass slime mold.

What about consciousness? Philosophers debate whether consciousness is special to the wet, organic brains of humans, apes, and dogs. Perhaps electronic intelligences, while their intellect may seem superhuman, will lack self-awareness or inner life. Or is consciousness emergent in any sufficiently complex network? This question crucially affects how we react to the far-future scenario that has been sketched. For instance, if machines are zombies, then we would not equate their experiences with ours, as the posthuman future would seem bleak. However, if they are conscious, then we should welcome the prospect of their future hegemony.

Additionally, what will be their motivation if they become fully autonomous entities? There is no way of knowing. Recall the variety of bizarre motives—ideological, financial, and religious—that have driven human endeavors in the past. Recall the aliens in popular science fiction novels. They could be contemplative, realizing that it is easier to think at low temperatures—far away from any star, or hibernating for billions of years, until the microwave background cools below three degrees kelvin. However, they could be expansionist. This seems to be the expectation of most who have contemplated the future trajectory of civilizations.

Assuming that life originated on Earth alone, it need not remain a trivial feature of the cosmos; hypothetically, humans could prompt a diaspora, whereby increasingly complex intelligence spreads throughout the galaxy, transcending our limitations. Also, the "sphere of influence," or "frontier of conquest" could encompass the whole galaxy; thus, spreading via self-reproducing machines, transmitting DNA and instructions for 3-D printers, etc. The leap to neighboring stars is an early step in this process; interstellar voyages, or intergalactic voyages, would hold no terrors for near-immortals.

Moreover, even if the only propellants utilized were the ones we currently know, galactic colonization would take less time—measured from today—than elapsed time since the Precambrian explosion as well as the emergence of primates, if it proceeds.

Consequently, the expansionist scenario could potentially make our descendants more conspicuous, allowing alien life forms to become increasingly aware of our existence. However, there is a question we often find ourselves asking: Is extraterrestrial life already out there? Are there other "expansionists" whose domain may affect our own?

There is not an answer to these questions. The emergence of intelligent life may require a rare chain of events, suggesting that this event has not occurred anywhere else. Such revelations would undoubtedly disappoint SETI researchers; however, suppose we are not alone. What evidence would we expect to find?

Assume that life began on multiple planets, and that on some of them, Darwinian evolution followed a similar track to what has happened on Earth. Still, it is highly unlikely that the key stages would be synchronized. If the emergence of intelligence and technology on another planet lags significantly behind Earth, then that planet would reveal no evidence of extraterrestrial life. Yet, on a star that predates the Sun, life could have had a head start of a billion years or more.

The history of human technological civilization is measured in millennia. So, it may be at least two more centuries before humans are overtaken or transcended by inorganic intelligence, which will then persist, continuing to evolve on a faster-than-Darwinian timescale, for billions of years. Organic human-level intelligence will be a brief interlude before machines take over. So, if alien intelligence had evolved similarly, then it would be unlikely to discover it while embodied in that form. If extraterrestrial life were detected, acquiring electronic signals would be more likely than discovering creatures that are neither flesh and blood, nor on other planets.

Moving forward, the Drake equation must be reinterpreted. The lifetime of an organic civilization may be a millennium at most; however, its electronic diaspora could continue for billions of years. Moreover, if SETI does succeed, then it is doubtful that the signal received would be a decodable message. Such an event would likely reveal a byproduct, or malfunction, of a complex machine that is far beyond our comprehension.

Incidentally, referring to them as alien civilizations may be too restrictive. A civilization connotes a society of individuals; in contrast, extraterrestrial life could potentially be a single integrated intelligence. So, even if messages were being transmitted, we may not recognize them as artificial because it would be difficult to decode them. A veteran radio engineer familiar only with amplitude-modulation might have a hard time decoding modern, wireless communications. Indeed, compression techniques aim to make the signal as close to noise as possible; insofar as a signal is predictable, then there is scope for more compression.

Moreover, SETI's focus has been on the radio aspect of the spectrum. Naturally, in our state of ignorance, we should thoroughly explore all wavebands: the optical and X-ray band. We should also remain aware of further evidence indicating non-natural phenomena or activity. One might seek evidence of artificially created molecules, such as chlorofluorocarbons (CFCs), in an exoplanet atmosphere, or massive artifacts, such as a Dyson sphere. It may even be worth seeking artifacts within our solar system. Perhaps we can avert visits from aliens; however, if their civilization mastered nanotechnology and transferred its intelligence to machines, the invasion might consist of a swarm of microscopic probes that could have evaded notice. In this regard, it would be beneficial to surveil the sky for especially shiny, or oddly shaped objects lurking among the asteroids. Ultimately, it would be easier to send a radio or laser signal, rather than traversing the spectacular distances of interstellar space.

Finally, a few speculations regarding the far-distant future—beyond the evolutionary stage any alien could have reached in the present universe, even with maximal advantage. Fast forward to an astronomical timescale, millions of years into the future. The ecology of stellar births and deaths in our galaxy will proceed more and more slowly, until eventually jolted by the environmental shock of an impact with Andromeda—four billion years hence. The debris of our galaxy, Andromeda, and their smaller companions within the local group will thereafter aggregate into one amorphous galaxy. Distant galaxies will move farther away, receding rapidly until they disappear. However, the remnants of our Local Group could continue for far longer—time enough, perhaps, for the Kardashev Type III phenomenon to emerge as the culmination of the long-term trend for living systems to gain complexity and negative entropy (Freeman Dyson, Seth Lloyd, and Lawrence Krauss have written on this topic).

Yet, there are limitations set by fundamental physics: the number of accessible protons (which can transform into any element) and the amount of available energy, which can also be transformed into various forms. Atoms that were once trapped in stars and gas could be transformed into structures as intricate as a living organism or a silicon chip, but on a cosmic scale. This process, though, would require a node and lattice architecture to minimize the time lags caused by the speed of light. Some science fiction authors anticipate extensive engineering to create black holes and wormholes. In other words, there are concepts far beyond any technological capability that we can conceive, but not in violation of the aforementioned physical laws. Furthermore, the limit exceeds the mass energy in the Local Group. So, it is consistent with physical laws to marshal receding galaxies before they accelerate over the horizon, drawing them in—thus, creating a segment of an Einstein static universe with a mean density in which cosmic repulsion (lambda) balances gravity.

The preceding points are consistent with the laws of physics as well as the cosmological models as we understand them, assuming that the force causing cosmic acceleration persists (e.g. dark energy or Einstein's lambda). Nevertheless, we should accept that there is much we do not understand. Human brains have changed little since our ancestors roamed the African savannah and coped with the challenges that life then presented. It is remarkable that brains have allowed us to make sense of the quantum as well as the cosmos, distant from the world in which we evolved. Scientific frontiers are advancing fast and, although we may come to a stalemate at times, one day the answer to life's many questions will come into focus. For instance, there may be phenomena, crucial to our long-term destiny, of which we are not yet aware. Physical reality could encompass complexities that neither our intellect, nor our senses can grasp. Some electronic brains may have a different perception of reality; therefore, we cannot predict or understand their motives. Hence, our inability to assess whether their silence signifies their absence or their preference.

Conjectures about advanced or intelligent life are more unpredictable than those regarding simple life. This suggests three things about the entities that SETI searches could reveal: They will not be "organic" or biological, they will not remain on the planet where their biological precursors lived, and we lack the ability to fathom their intentions.

Even these speculations fail to take us to the utter limits. I have assumed that the universe itself will expand at a rate that no future entities have power to alter. And that everything is, in principle, understandable as a manifestation of the basic laws, governing particles, space, and time that have been disclosed by contemporary science. So, are there new laws awaiting discovery? Will the present laws be immutable, even to a Type III intelligence that can draw from galactic-scale resources?

Post-human intelligences (autonomously evolving artifacts) will achieve the processing power to simulate living things—even entire worlds. These super- or hyper-computers would have the capacity to simulate not just a simple part of reality, but a large fraction of an entire universe.

Then, the question arises: If these simulations exist in far larger numbers than the universe themselves, could we be in one of them? Could we ourselves not be part of what we think of as bedrock physical reality? Could we be ideas in the mind of some supreme being who is running a simulation? Indeed, if the simulations outnumber the universes, as they would if one universe contained many computers making many simulations, then the likelihood is that we are the artificial life, in this sense. This concept opens up the possibility of a new kind of "virtual time travel," because the advanced beings creating the simulation can, in effect, rerun the past. It is not a time-loop in a traditional sense; it is a reconstruction of the past allowing advanced beings to explore their history.

The possibility that we are creations of some supreme (or super) being blurs the boundary between physics and idealist philosophy, between the natural and the supernatural. We may be in the matrix rather than directly manifesting the basic physical laws.

What, then, does this mean for us as we approach the middle of the twenty-first century? Many of the questions I've posed are simply unanswerable from our vantage point. But it is clear that advancements in biotech, artificial intelligence (and the human/machine interface), and space technology will shape the future of both humanity and sentient life in our corner of the universe. The future is truly ours.

Stella Infantes

Kacey Ezell and Philip Wohlrab

When she was little, Kacey Ezell dreamed of being an astronaut or a dragonrider. She became a helicopter pilot instead—which is pretty close. Kacey writes SF, mil-SF, fantasy, alternate history, and horror fiction for Baen and Chris Kennedy Publishing. She is married with two daughters and spends her time flying, writing, or being a cheer mom.

Philip Wohlrab is a military medic with more than a decade of experience practicing in combat medicine, advanced lifesaving, and basic lifesaving skills, both on and off the battlefield. He is a senior instructor in the combat medic sustainment program for the Virginia Army National Guard and is a senior field sanitation instructor. He earned a masters of public health with a focus on disease prevention. When not working for the Army he, of course, writes science fiction.

Dr. Maureen Fitzberger, "Mo" to her friends, stared at the readout on the screen. She realized her mouth was open and closed it with an audible snap before anyone could happen by her little lab cubicle and catch her staring like an idiot. She swallowed hard, trying to ignore the sudden thumping of her heart as she glanced around, and then looked at the screen again.

"...*viable embryo. Gestational age: twelve weeks...*"

It *wasn't* possible! Well, it *shouldn't* be possible. Sure, she'd been taking the hormone supplements to extend the lifespan of her fertility like every other female member of the crew, but just like all of them, she also had a birth control implant! What the hell had happened?

Okay, so she knew *what* had happened. That lovely young pilot named Haskins. But that had just been fun, blowing off steam. Mo hadn't seriously been considering him as a permanent partner!

"What's going on, Mo?"

Mo looked up as her medic walked in. She forced her lips to smile and hoped she didn't look whiter than her normal redhead pallor. She certainly felt pale.

"Everything okay?"

"Yeah, Doc," Mo said, shaking her head slightly and forcing herself to focus on her friend's face. Desmond "Doc" Joel wasn't actually a medical doctor, but he'd earned the nickname as a military medic back on Earth. After he'd gotten out, he'd been certified as a civilian paramedic and eventually found his way into the pool of applicants selected for this scout-colony mission.

The irony, of course, was that despite her years in applied genetics, Mo *was* trained as an M.D. Like everyone else, though, she called the medic "Doc." Military nicknames were like that. They snuck in everywhere.

"Something's got you rattled," Doc insisted, setting down the customary coffee thermos and lowering his muscled body into the chair at his usual workstation next to hers. "You look like you've seen a ghost."

"The opposite, actually," Mo heard herself say, then gave a kind of strangled laugh. "I suppose you'll have to know eventually. I certainly won't be able to do what's necessary without you."

"What?" Doc asked, his smile draining away. His blue eyes went flat and calm in a way that Mo had only seen a few other times, usually when a patient wasn't going to make it. It was the look of a professional compartmentalizing their feelings and preparing to do the job, whatever it took. Mo didn't know much about Doc's military service, but she knew he'd seen, and presumably done, some heavy stuff. Most people didn't volunteer for a one-way mission like theirs unless they had.

"No," Mo said quickly, swiping her screen to send her display over to his workstation. "Nothing like that. It's just . . . something unexpected has developed. My implant failed, apparently. I'm twelve weeks pregnant."

"Pregnant," Doc said, and Mo could hear a breath of relief in his tone. "Damn, Mo, you scared me. But...is that good?"

"I don't know," she said, fighting the urge to bite her lip. "I mean, I suppose so... It means the hormone therapies are working. But we're just not set up to provide prenatal care here!"

"Yeah, but how hard can it be?" Doc asked, giving her his usual quick grin. "I mean, women had babies for millennia on Earth before the invention of the o-b-g-y-n, right?"

"Yeah, and had an astronomical mortality rate for mother and child!" Mo said, unable to keep from rolling her eyes. Doc's smile grew, indicating that he'd tossed the banter her way deliberately—probably to help her relax.

He was good, real good. She let out a little laugh and shook her head again.

"All right," she said. "Point taken. It's not like I don't know what to do, after all. And you can deliver the baby when it comes to that. But there's more to think about before we get to that point."

"Like what?" Doc asked. He leaned forward and began tapping on his keyboard.

"Well," Mo said, "lots of things. All those millennia of women weren't part of an interstellar scout-colony expedition. Gestational experiments in space have had...mixed results."

"Yeah," Doc said. "Microgravity and radiation, right? The kid's gonna need gravity to be able to develop an inner ear, and you're gonna need to stay in the shielded part of the ship for the duration...unless...do you want me to put you back in stasis? I can wake up one of the other docs to take your place on this last shift, then wake you up right before we make planetfall."

Mo shook her head.

"No," she said, "that's no good. One of the stasis prep drugs is contraindicated for pregnancy. Interferes with early neurological development. In trials, most mammalian mothers spontaneously aborted, and those who made it to term didn't produce viable offspring. We can't put someone in stasis until they're over five years of age."

"And we make landfall in three years," Doc finished for her. "Okay, so no stasis."

"No stasis...but there's another thing."

"What's that?" Doc asked.

Mo leaned toward him and lowered her voice.

"Doc," she said, "this...this is an opportunity."

"What do you mean?"

"Well," she said. "We don't know much about our new planet, right? I mean, we have the survey reports, but they're several centuries old by now. And I know that's nothing in geological time, but still...that's why we're a scout colony. Scouting it out first, establishing a beachhead for others to follow, right? What if we find conditions that require adaptation?"

"We're expecting that, aren't we?" Doc said, his brow furrowing slightly in a frown. "That's why we sent the high-speed probes ahead to Bonfils well before our ship launched. As we get closer and receive more data, we'll compensate with mechanical means first, and then follow up with gene therapies as necessary."

"Right," Mo said. "But what if we could start right away with the gene therapies? What if we could use my baby's stem cells to create genetic solutions before we even make planetfall? Those of us already born could see moderate adaptive success, but the next generation...the next generation could be fully adapted! No mechanical assistive devices required, regardless of the air composition, mineral cocktail, weird alien bacteriological or viral beasties...all of that! We can start right away!"

As she spoke, the idea took root in Mo's mind, and the possibilities started to open up one by one. She could hear her voice accelerate as she got more and more excited about the idea, and she felt her cheeks heat up with an excited flush that chased away the last of her surprise pallor.

"So, you'd use your child as a guinea pig?" Doc asked, arching an eyebrow.

"Oh, bugger off," Mo said, laughing. That was Doc, always with the banter. "Yeah. Yeah, I would, if it meant that she had a better shot at a quality life on her new home. We're explorers, Doc! They told us that over and over again when we prepped for this mission. Hell, it's *why I'm here!* I'm supposed to design gene therapies to help *Homo sapiens* become successful on this new planet. But what if what we need is to *stop* being *Homo sapiens* altogether? What if what we need is to become something else? What if we need to become *Homo stellaris*?"

Mo heaved, but she'd already launched her meager breakfast into the toilet chute and so all that came up was a thin stream of bile. Her stomach roiled again as she closed her eyes and prayed to whomever may be listening to make it stop.

"Here," Doc said. He sounded supremely sympathetic. Mo felt something cold and wet settle against the back of her neck. "It doesn't help, but it feels nice and cool."

"Thanks," Mo croaked, spitting. She let out a sigh and settled back onto her heels from her kneeling position. "Morning sickness blows."

"Technically, it's midafternoon," Doc said.

In lieu of an answer, Mo lifted her left hand vaguely in his direction, middle finger extended. Doc let out a gravelly chuckle in response.

"Just trying to take your mind off your stomach."

"Can we talk about something else?" Mo reached up and pulled the wet cloth off the back of her neck. Doc was right, it had felt nice.

"Sure. Wanna see the results of your last ultrasound?"

"Okay" Mo said, straightening slowly up from her lean. When her stomach didn't protest too much, she rolled back over her feet until she was sitting on her butt. Then she swiveled around to lean her back against the bulkhead. Without being asked, Doc held out a bottle of water to her. She grasped it with a grunt of thanks and took a mouthful. It tasted like Doc had added peppermint oil, and probably some electrolytes. She washed it around and then spat it carefully in the toilet chute, followed by a small sip.

So far, so good.

"Okay," Doc said, once it became clear that Mo wasn't going to start throwing up again. "Everything seems to be proceeding apace. Fetal development is meeting the milestones you outlined for the most part. That's the good news."

"What's the bad news?"

"The bad news is that I followed a hunch and did some research. I've figured out why your morning sickness is so bad."

"Wait . . . what? How's that bad news?"

"Because," Doc said, leaning back on his desk and crossing his arms over his chest. "It's the anti-spin cocktail for your inner ear that's exacerbating the problem. Do you remember being briefed on the effects of simulating gravity using rotation?"

"Yes," Mo said, swallowing hard against another wave of the ever-present nausea. "Turn your head and you puke."

"Succinct," Doc replied with a tiny grin. "But essentially correct. So we give everyone a med cocktail to mitigate the effects on the inner ear that cause the disorientation and nausea. Problem is . . . here, can you read this?"

He held out a tablet. The display showed a journal page. With a tap of his finger, Doc highlighted a few words in a paragraph about midway down.

" . . . such a cocktail is contraindicated for pregnancy. Test subjects in all stages of gravidity showed an elevated incidence of hyperemesis gravidarum. Though no other side effects were noted, offspring of mothers experiencing hyperemesis showed markedly lower birth weights and correspondingly higher incidences of birth complications . . ." Mo looked up from the slate. "So the drugs are making it worse."

"They are," Doc said. "We need to get you off of them, because while you're meeting milestones, you're doing so slower than you should. That means that you're going to have a hell of a time in the living quarters. Though, if my gut is correct, your inner ear will eventually be able to adapt, and you'll get used to it, somewhat. Ever heard of a Bárány chair?"

"No," Mo said, once again fighting nausea.

"They used them in the old days to cure pilots of airsickness. Basically, it's a chair on casters. The student pilots would sit in them, spin around, tilt, and have to do tasks until they felt like throwing up. Eventually, the inner ear adjusts and the student doesn't get sick anymore. It worked really well. I suspect this will be similar."

"How do you know this stuff?" Mo asked, lifting her head and trying not to glare at Doc as her stomach started to roil again.

"Grandma Ida was a pilot," Doc said with a grin. "She had to spin all the way through her primary training. But I'll tell you, the woman never once got motion sickness. Not ever."

"Okay. I come off the drugs, and maybe start gaining some weight. But I'll still throw up occasionally because of the spin-induced vertigo."

"Until your inner ear adjusts, probably."

"Fantastic," Mo groaned, dropping her forehead to her knees once more.

"We still need to monitor your radiation exposure too," Doc said. "Your lab and living quarters are pretty well shielded, but the problem with being on the end of the spin arm is that it's not as well shielded as the center of the ship...but I think it's still worth it to have normal skeletal and inner-ear development. That said, if the cosmic ray detector goes off then you need to hightail it back to one of the radiation shelters."

"Agree," Mo said, "and I have an idea about the microgravity problem."

"Oh?"

"Yeah. There was a study back in the early 2000s. It's pretty obscure now, but I found it when I was doing my research. Basically, a bunch of snails—several animals, really, but the snails are the important ones here—were gestated in microgravity. When they were hatched and returned to Earth, the animals had a hard time establishing a sense of balance. Like we've talked about before."

"Right," Doc said. "And the further trials involving placental mammals bore that out with underdeveloped inner-ear mechanisms. The offspring eventually adjusted, but it took longer than the period of readjustment for adults."

"Yes, but not for the snails," Mo said, managing a wan smile, even though it felt like her insides were about to crawl out of her mouth again. "*They* adjusted quicker than any other space-born animal, but that detail was never emphasized in the larger study. But when I looked into it, I found out why. Snails have the same kind of gravitoreceptors as humans do. Ours is in our inner ear. Theirs is...well, elsewhere, anyway. The point is, the space-born snails had gravitoreceptor sizes that nearly *tripled* the norm."

"They were compensating," Doc said with a small grin.

"Yes," Mo said. "Exactly. So while I still want to stay out of microgravity for the skeletal development, I can develop a gene therapy regimen that will grow my baby's gravitoreceptors similar to those of the snails...so *she* won't have to puke her guts out if she has to live in space."

"Nice work," Doc said. "If you're sure..."

"I am," Mo said. "Now if you'll excuse me, I'm going to throw up again."

"Damn it," Doc said. He moved the ultrasound wand slightly, but the image on the display didn't change.

"Hello little hamburger," Mo said, feeling giddy and silly. "Read it and weep, Doc. You lose."

"Fine, a bottle of my best hooch," the medic said. "Too bad you can't drink. I guess I'll just have to drink it for you. How did you know?"

"That she's a girl? I don't know," Mo said. "I just knew. Gut feeling, I guess. Some mystical mother-child connection bullshit, perhaps?"

"Huh, maybe," Doc said. "But you're right. She's definitely a girl. And all her development looks good, if I'm doing this right."

"Looks like you are to me," Mo said. "The big thing is moving the wand slowly, understanding that you're seeing in layers, but yeah. I agree. She's doing just fine, aren't you, little star?"

Doc smiled slightly and ducked his head, likely at the soft tone that crept into Mo's voice when she addressed her unborn daughter. Mo figured she'd catch hell for it from him later, but she honestly didn't care. Pregnancy was affecting her in ways she hadn't ever imagined it would. Soft, vulnerable emotions welled up within her without notice, filling her eyes and making her sniffle. Yet, a hard practicality had solidified within her mind: She would protect the precious life inside her from whatever loomed on their new home. She would manipulate her daughter's genome as much as it took to ensure that she would not only survive colonizing a new planet...but thrive.

"Are you going to cry again?" Doc asked. Mo blinked suddenly and forced her eyes to focus on him.

"Oh, shut up, sore loser," she said gruffly. Doc just chuckled and handed her a large alcohol wipe.

"Clean yourself up," he said. "The all-hands meeting is in a few minutes. You're still sure you want to go through with this?"

"Gotta tell them all sometime," Mo said. "Might as well get it over with. Especially since I'm asking them to change our mission priorities once we make landfall."

"Fair enough," Doc said. "She's your baby."

"That's just it," Mo said as she sat up and began cleaning the ultrasound gel off of her barely rounded stomach. "She is, but in a way, she will soon belong to all of us."

"...as we draw nearer to our new home, we're able to receive better and more complete intel reports on conditions at the surface. I will use that data to determine a course of prenatal and early childhood gene therapies to maximize my daughter's chance of successful adaptation to the new world. She will be, in effect, the mother of a new race of humanity. She will be this planet's Eve."

Mo took a breath and looked around the conference room at the stunned faces. Silence greeted the end of her prepared remarks as the other members of the crew digested the information she'd just given them. Nerves began to flutter in her stomach, and she glanced over at Doc, who leaned against the entrance hatch.

He flashed her a thumbs-up and gave her a nod, but it didn't help. If only someone would *say* something...

The shift captain looked around and slowly raised his hand.

"Yes!" Mo said, pointing at him like a drowning woman points at a life preserver. "Captain Johnson, you have a question?"

"Well," the captain said, coming to his feet. "Yes. Several, in fact. But first, I want to offer my congratulations, Mo. I know this wasn't planned, but I have always thought that a new baby should be celebrated, regardless of the circumstances."

"Thank you, sir," Mo said, giving him a tremulous smile. She gripped the edge of her podium and willed her knees to stop shaking.

"Second," the captain said. "I just want to be sure I understand what you're telling us. You propose to create a course of genetic therapy to ensure your daughter's full adaptation to our new planet...and what? To extend that therapy course to *all* of the children that we will hopefully, eventually have?"

"Yes, sir," Mo said. "Or at least to all whose parents want it for them. We know that Bonfils is a close Earth analogue, but not exact. Some adaptation will be required. Our original plans were to compensate with mechanical means as long as necessary...but if I can give my daughter a childhood free of breathing masks and other assistive devices we might need, I would prefer that option. Please note, I mean no insult to those who might choose otherwise for their children. I just... I'm offering the option. The adaptations will have to be made somehow. With this pregnancy, we have the option to make them sooner rather than later."

"Understood," Captain Johnson said. "Next question: You mentioned a proposed change in our mission timeline. Could you elaborate on that?"

"Of course, sir," Mo said. This was the part she'd been dreading. Captain Johnson had been chosen to lead the scout-colony team during the last five years of their journey, plus the landfall and initial setup phases of the mission. Until the colony held its first planet-side elections, he was responsible for their success or failure. Naturally he'd want to know why she was proposing to change things around.

"Our current timeline calls for the first generation to be born planet-side approximately ten Earth years—or eight local years—after landfall. The plan is to give the team time to meet scouting objectives and ensure that the colony is securely established before bringing new humans into the mix."

"Two objectives which are still necessary to meet, don't you think, Dr. Fitzberger?" the captain asked, looking at her with his eyebrows raised behind his rimless spectacles. He folded his arms and waited for her reply. Not a good sign, as far as Mo was concerned.

"Yes sir," she said. "No question. But the colony's *ultimate* objective is to exist, right? I mean, if you'll forgive the misquote, we're under orders to 'multiply and replenish' as well. That's the whole point, isn't it? The colony will fail if we aren't diligent in seeing to that first biological duty."

"You propose that we prioritize breeding over the other necessary tasks involved in creating a livable habitat for ourselves?"

"Not at all, sir," Mo said quickly. She liked and respected Captain Johnson, but a spark of irritation flared inside her, steadying her nerves. "But let's look at the data in front of us. All indications are that my implant failed in part because of the fertility-extending cocktail I've been taking. That we've all been taking. Chances are that as we go on, other implants will fail as well. My baby won't be the last happy accident we're going to see. Especially since the male implants are known to have a failure rate of over twenty percent. If our fertility extensions have weakened the female implants, too...

"I'm not arguing that we don't need to proceed with caution. We will still need to engage in scouting and establishment procedures. But, I would argue that instead of holding the majority of our crew members in stasis while these procedures are carried out, we can begin awakening them on an accelerated timetable. We can double our workforce, and the women who are so inclined can have their implants deactivated and begin the work of creating the first generation immediately."

"And how do you propose to feed double the workforce, plus pregnant mothers?"

Mo smiled.

"Captain, I'm a genetic engineer. You give me a food source and a few months, and I'll have a caloric surplus of food for you in short order."

"And this is all so that your child can have peers to interact with? Don't you think that's spectacularly selfish?" Mo turned to look at the voice of the woman who spoke, and didn't recognize her. She sat in the back of the room, arms and legs crossed, her face set in lines of disapproval and her eyes radiating scorn.

Mo drew in a deep breath and looked around the room at the other faces watching her. She met Doc's eyes, and he gave her a little smile and a nod. So, she squared her shoulders, and turned back to address the man who commanded the ship.

Okay, Mo, she told herself. *Time to sell it.*

"Captain, I just found out that I'm to be a mother rather sooner than expected. Apparently, it's a survival instinct of our species that mothers will do *anything* to assure the welfare of their children. So yes, I do believe that my baby needs other children to interact with, and yes, that was the reason I started coming up with this plan . . . but the truth is, if you look at it . . . it's good for the colony, too. Starting the next generation early, especially if we manipulate their genomes for maximum adaptability, will only ensure a stronger, deeper foothold on Bonfils. That, sir, is the very definition of a win-win situation."

"Captain."

Another woman in the back of the room stood up, breaking Mo's concentration. She turned to look at this new voice—one she recognized, this time—as her knees began once more to tremble.

"I'd volunteer to be one of the early mothers," the woman said. "My secondary specialty is in early childhood development, and so I could help with setting up appropriate curricula as soon as we have structures built on the surface."

"But what of your primary specialty, Rita?" the captain asked with a frown.

Rita grinned. "Food services. I'm a chef, Captain. So, when Dr. Fitzberger engineers her superfood, I'll figure out twenty different ways to serve it to increase palatability."

"While raising the colony's first generation of children?"

"To a degree, yes," Rita said.

"Captain." That was Doc, who stepped away from the wall. "May I add something?"

"Medic Joel. By all means," the captain said, waving a hand.

"Dr. Fitzberger's plan is solid, especially if we accelerate the revival timetable for our team members currently in stasis, matching it against our available and increasing food supplies. If they're all awake shortly after landfall, then we have a labor force ready to go with more than enough hands to take shifts...thus allowing those who want to participate in the doctor's breeding program to do so."

"I wouldn't call it a 'breeding program,'" Mo said, nerves bubbling out of her in a laugh that rippled through the crowd, too. "I'm not pairing individuals up or anything like that. I'd just advocate that those who feel inclined to start early be allowed to do so...and I'll create the genetic enhancements that will ease adaptation for this first generation."

"What adaptations?" the captain asked.

"I've listed a few of them on the slide here," Mo said, as she flicked her wrist and a display screen came to life behind her head. "Slight changes to the red blood cells to compensate for Bonfils's lower oh-two saturation; more melanocytes to deal with the stronger UV radiation we've seen; increased size and sensitivity of gravitoreceptors to enhance adaptability to micro-gravity; increased bone density to compensate for slightly higher gravity...things like that. We'll know more about our specific requirements as we continue to get more and better data from the advanced probes."

The captain paused and looked around the room. Mo took a moment to do the same, and what she saw warmed her. People were nodding and smiling; a few couples were talking animatedly in close whispers. Most seemed in favor of her idea. She felt herself smile, and something eased in her chest.

"Well," the captain said. "I will take this under advisement. Let's recess this meeting for now. I'll give you all my decision tomorrow."

Though they weren't a military ship, per se, some customs and courtesies had wormed their way into the scout colony's culture. The rest of the room came to their feet as Captain Johnson turned and left the conference room. The moment the hatch closed behind him, the room broke out into the animated babble of too many conversations at once.

"Good job," Doc said as he pushed through the crowd to

approach Mo. "I think you've convinced him. Especially the bit about your adaptations. That was always going to be the danger with having babies out here."

"I can't believe we didn't think of it before," Mo said.

"Who says we didn't? You said it yourself... we didn't really have the real-time info we need. Now we do. Or we will, as soon as we start intercepting the data packets from the recon drones that were launched well before we took off."

"Speaking of which, I heard we're expecting to intercept the first burst of information in a day or two," Mo said. "We'll have that info and more as we get closer and perform more intercepts."

"Well, I suppose you'd better get ready to get to work, then!"

The information contained in the first recon probe data burst was encouraging. Bonfils showed a marked resemblance to Earth in the chemical composition of its air and soil, so no breathing masks would be needed after all. The plant life collected by the probe's bots showed higher traces of boron than Terran plants, as well as a few other interesting minerals.

Perhaps the most interesting thing wasn't totally news at all. The original remote surveys of Bonfils had noted some interesting coloration patterns on both flora and fauna that weren't visible to the normal human eye. However, they were clearly visible in the near-IR spectrum. The probe's data confirmed this trend and indicated that the near-IR coloration patterns were even more widespread than originally suspected.

"You should make her see in the dark," Doc said at one point, about five days after the probe's information returned.

"What do you mean?" Mo asked.

"I mean, you should give your baby the ability to see in the near-IR spectrum at night. Modify her rod cells. You can do that, can't you?"

"Sure, but why? I mean, we've talked about doing it for military applications, but adult gene therapies don't have as high a success rate. If you're going to modify DNA, it's best to do it while the patient is young... and not a lot of infants are showing interest in becoming high-speed tactical operator types."

"Because of this coloration trend," Doc said. He swiveled his display and magnified it so that Mo could clearly see the paragraph discussing the prevalence of near-IR fluorescent pigments on the surface. "It looks like most of the native life has it. That means it probably serves as camouflage or as a warning function. In either case, don't you think she'd find it useful to see?"

"Hmmm . . . you know, that's not a terrible idea," Mo said. "Problem is, if I do that, she'll never be able to turn her 'night vision' off."

"Well, sure she will. Near-IR is just light like the rest of it, right? She'll just need a tighter dark room if she wants total darkness."

"Have you got any idea how hard that will be to manage?"

"No," Doc said, "but I do know how easy it is to adapt to sleeping in the middle of the day. Had to do it all the time in the Army. It's your call, obviously, but it seems to me that this is a perfect example of using a mechanical adaptation—NVGs— when a genetic one would be better."

Mo pursed her lips and considered the question.

"Let me run some initial calculations and see just how we'd create the protocol," she said, "and we'll go from there."

In the end, she included the retinal modification as well. Because Doc was right. That sort of adaptation was exactly what she had in mind. *Homo sapiens* wasn't native to Bonfils. *Homo stellaris* would be.

As the ship drew closer to their new home, Mo found herself suddenly very popular among the other female members of the shift. One of the selection requirements for the scout-colony personnel had been a stated willingness to contribute to the next generation in the form of children. Most of the women had at least a little bit of curiosity about the process, and so they sought Mo out and questioned her closely as her body grew rounder and rounder.

"I swear," Mo joked to Rita as she lowered herself into a seat in the galley. "If I'm not careful, Captain Johnson's going to launch probes at *me*."

Rita laughed and plunked a cutting board with a handful of hydroponically grown zucchini down on the small prep counter in front of Mo. Chef Rita's rules were clear: You sit in her galley, you helped prep meals. Mo didn't mind. It gave her something to do with the restlessness that never seemed to go away.

"At least you're not losing your lunch on the hour anymore," Rita said. "I was starting to feel insulted."

"Ugh, don't remind me," Mo replied, picking up the knife. "Thinking about it still isn't a good idea."

"Fair enough. How close now?"

"Maybe another thirty days?"

"Are you ready?"

"As I'll ever be," Mo said. "I just want to be able to move without feeling like I have a bigger turn radius than this ship."

"I didn't think this ship had much of a turn radius," Rita said.

"Exactly."

"Is Pretty Boy excited?" the Rita asked. "I heard some of the crew got him good and wasted at the ship's baby shower we threw for you guys."

"I think he is," Mo said. "I don't think we're going to be compatible for a long-term relationship, but he seems to be pretty committed to co-parenting. He's been in on all of the gene therapy decisions we've made."

"And what have those been, if you don't mind me asking?" Rita looked up from where she was tying some kind of vat-grown roast together, interest sharp in her eyes.

"Not at all," Mo replied as she lined up the vegetables and began cutting. "I'm giving her near-IR night vision, in order to help her see the local flora and fauna cues that the probes found. I've made the gravitoreceptors in her inner ear larger, so that she'll be more sensitive to changes in gravity, and better able to orient herself in micrograv. I've changed her digestion slightly, so that she'll be better able to metabolize the boron-heavy food sources available. That's most of it, honestly . . . oh! And I went ahead and shifted her fertility to later in her life. I always thought it ridiculous that we women are capable of bearing children when we're still mentally and emotionally children ourselves. So she'll experience menarche sometime between sixteen and eighteen, and remain fertile until menopause at sixty."

"Oh, that's smart," Rita said.

"I hope so," Mo said. "There were some very good evolutionary reasons for humans to be able to bear children young . . . but my hope is that we're able to compensate for those reasons in other ways."

"Technology?"

"Mostly."

"You're saying it's a gamble," Rita said.

"This whole thing is a gamble," Mo pointed out.

"True enough—these adaptations will also be applied to our babies? Those of us who will start bearing on the surface?"

"Only if you want," Mo said. "If not enough couples choose them for their children, they'll be bred out of the genome. But if it turns out that they're useful . . . well . . . then they'll be naturally selected."

"It sounds cold when you say it like that."

"I know," Mo said with a sigh. "But it's true."

"Well, fair enough. Here, let me have that zucchini."

At thirty-eight weeks, Mo's blood pressure began to climb.

"I don't like this," Doc said as he looked over the display of her vitals. "Your hypertension is trending upward. I think we might want to think about getting the kid out."

"You mean inducing?" Mo said with a grimace as she lay on the exam table in the ship's sickbay. "I really don't want to do that."

"I don't think you're going to have much choice," Doc said. "Rita's been watching your diet like a hawk, and still your BP numbers are up. Make the smart call, doctor."

"You're right," Mo said, sighing. "I just . . . damn. I hate Pitocin. It's criminal that modern medicine *still* hasn't found a better method of labor induction than to pump mothers full of that nasty stuff."

Doc bit his lip, looked at the display, and then over to Mo.

"There's another option," he said softly, sounding tentative. Mo blinked and looked sharply at his face. "Tentative" wasn't Doc's usual style.

"What do you mean?"

"I could do a laser C-section," he said. "We were trained on it for emergencies, and I've been studying up . . . just in case."

Mo drew in a deep breath. She hadn't considered the possibility of a C-section. Normally, a surgical procedure like that would be beyond a medic's scope, except for emergencies. But the ship's sickbay had automated anesthesia capabilities, to include spinal blocks, so she could supervise and direct. *She* hadn't done a C-section since she was a resident . . . but it was actually a pretty simple procedure, when it came right down to it.

"Are you up for that?" she asked. "I don't want to push you too far outside of your comfort zone, but it *would* eliminate a lot of variables in the birthing process."

"I am," Doc said, "if you trust me, and if you're willing to oversee it."

"Doc," Mo said, "I trust you with my life, and my baby's life. If that's not clear by now, you're the dumbest SOB I've ever met."

Doc let out a laugh, and just like that, his habitual confidence returned. Mo was glad to see it.

"All right," he said. "When do you want to do this?"

"Now," Mo said. Certainty settled over her mind as a sudden surge of energy rippled down her nerves. She pushed up to a seated position on the exam table. "Right now."

"Are you serious?"

"Why not? You said it yourself. My BP is only climbing. I'm at thirty-eight weeks, so she's close to full term. There's no real reason to delay. You go set up the anesthesia automation and I'll get hold of Haskins. He wants to be here when his daughter is born."

"Wow, you *are* serious. Okay...okay," Doc said, and stood up. He looked around wildly, and then sat back down at his display console while he collected himself.

"Relax, Doc," Mo said as she carefully stepped down from the exam table. "It's going to be fine. Just set up the anesthesia. I'll be right back."

Ten minutes later, Mo lay back on that same exam table as the anesthesia spread a cooling numbness from just below her breasts down to her feet.

"I can feel it," she said out loud.

"Good," Doc said, all trace of his earlier nerves gone as he focused on the display readout. "Looks good over here too. All the levels are showing optimal."

"Perfect," Mo said, and then turned her head to give Haskins a reassuring smile. He sat on a stool in the corner of the room, well out of the way. When the pilot arrived, Doc informed him in no uncertain terms that he was to sit there, not move, and not say a word until spoken to. The normally brash and cocky father-to-be agreed, his face paler than normal.

"All right, Mo," Doc said, pulling her attention back to the task at hand. "Are you ready?"

"Let's do this," Mo said. "Set your laser scalpel to two point one..."

✧   ✧   ✧

"She's the most beautiful thing I've ever seen in my entire life," Haskins said, his voice thick with awe. Mo smiled and tried not to resent him as he cradled the amazing, precious, overwhelmingly perfect little life that was their daughter. It wasn't that she begrudged him the time...no...no it was. Haskins was a good man, and would no doubt be a good father, but damn it, *she* wanted to hold the baby!

"All done here, Mo," Doc said, sounding both tired and triumphant. He had sutured her up himself, since the automated sickbay was busy keeping her anesthesia levels correct. "Just move slowly, okay?"

"I want to hold her," Mo said. She knew she sounded demanding and unreasonable, but at that moment, she didn't care. She'd carried the kid for nine months, she ought to be able to cuddle her!

"Sure, baby, sure," Haskins said, his voice breaking with joy as he carried his child over to her mother. "Here, she's right here. Look at her, Mo! Look what you made! She's even got your red hair."

He bent down and laid the blanket-wrapped bundle in Mo's arms. For the first time, Mo looked into the face of her daughter.

She was wrinkly and red, and looked thoroughly pissed off. But Haskins was right, she did have a shock of dark red hair on top of her head. Mo blinked as her eyes filled, and sniffled, then bent to press a kiss to the baby's forehead. Her lips had never touched anything so soft.

"Stella," Mo whispered. "Stella Dare Haskins."

"Oh, Mo," Haskins said, "You don't have to do that..."

"You're her father," Mo said. "She'll carry your name. Unless you object?"

"Hell no," Haskins said with a laugh. "Not in the least. I just figured you'd want to give her your name."

"That's okay..." Mo said. She would have said more, but she trailed off as Stella opened her eyes. From what Mo understood, most babies' eyes were a dark and indeterminate blue for several weeks after birth. Not Stella's. Her modified eyes gleamed green in the harsh light of the sickbay.

"Congratulations, you two," Doc said. "And welcome, Stella Dare. I would normally say welcome to Earth, but that doesn't exactly work here. So instead, I guess we'll go with this: Welcome to the Stars."

EPILOGUE

Doc picked up the new infant and rewrapped him in a bundle of swaddling. A healthy boy, he had been given all the advantages, (or upgrades as Doc still liked to call them), that had been worked out with Stella Dare. The infant looked up at him with eyes that were still an unnatural green to Doc's own Earth-born eyes, but would soon become the norm for Bonfils. The baby burbled at Doc as blue eyes peered intently into the boy's green.

"What sights will these eyes see? What new things will you discover?" Doc whispered to his son.

In response a new voice spoke up, a young child's voice.

"Can I see him now, Mr. Doc?" asked Stella Dare Haskins, the first *Homo stellaris*. Like Doc, her too-green eyes stayed riveted to the infant. "Please?"

"Sure, honey. You just have to sit down first."

After waiting to be sure she was settled in a chair, Doc squatted down to Stella Dare's level, and gingerly placed the baby boy into her waiting arms. She held him with all the solemnity of a child of four given a great responsibility. Doc still stayed crouched, ready to take the infant from her when needed.

Doc looked up as Mo walked into the room, having washed up from delivering the new child. He winked at Mo as she first looked at Stella and then to him.

"She was anxious to see him," he said, "and besides, it isn't every day you get to welcome the first child born planet-side, I bet she remembers this moment."

Mo smiled at Doc. "Sure," she said. "Mama's sleeping well. She'll be out for a few hours."

"Thank you," Doc said. "For everything."

"No thanks needed, Doc, you know that."

For her part, Stella Dare could at first only stare at the new boy, but at her mother's encouraging nod, she began to speak.

"Hi," she said. "My name is Stella. I've been waiting for you to come play with me. I can't wait until I can show you the garden, it's so big! We are going to have so much fun there. And then there's the heffalumps, you are going to like the heffalumps, they are my friends. They'll be your friends too, and then..."

Maintaining Crew Health and Mission Performance in Ventures Beyond Near-Earth Space

Mark Shelhamer

Dr. Shelhamer is on the faculty of Johns Hopkins where he started as a postdoctoral fellow in 1990. He has bachelor's and master's degrees in electrical engineering from Drexel University, and a doctoral degree in Biomedical Engineering from MIT. He had support from NASA to study various aspects of sensorimotor adaptation to spaceflight, amassing a fair amount of parabolic flight ("weightless") experience in the process. He also serves as an advisor to the commercial spaceflight industry on the research potential of suborbital spaceflight. From 2013 to 2016 he was on leave from his academic position to serve as Chief Scientist for the NASA Human Research Program at the Johnson Space Center. In this role, he oversaw NASA's research portfolio to maintain human health and performance in long-duration spaceflight.

Human spaceflight is challenging and calls on great personal resources even from professional astronauts. This has always been true, and its truth is not diminished by the fact that we now "routinely" send people into space for six months (or even a year) at a time on the International Space Station (ISS). A sense of the challenges in maintaining human presence in space can be gleaned from a glimpse at the physiological and psychological procedures (countermeasures) in place to maintain health and performance in this demanding environment. We will discuss those in a moment. When people venture further out, in distance and duration—to the Moon, to Mars (in the foreseeable future), and eventually beyond, these countermeasures will be taxed, in some cases beyond their limits. New interventions and conceptualizations will be needed: the enabling of the crew to deal, on its own, with issues that have not been identified before the mission begins. In other words, it is not the countermeasures themselves that will be as important as the *mission structure* that is put in place to create countermeasures "as needed" when new unexpected circumstances arise. We will discuss how this might be accomplished.

All of this is challenging enough, but when we then consider journeys not of *exploration* but of *settlement* or *colonization* to destinations that make Mars seem like a close neighbor, we enter a whole new realm of thinking. We assume that human physiology and psychology will be the same, and thus some of the same approaches to maintaining their integrity will still be relevant. However, the range of physical, mental, and emotional stressors will take on a new magnitude, leading to new problems that require new solutions. Extrapolations of solutions from today's flights will be stretched, hopefully not to the breaking point, but it is clear that new ways of thinking about health and performance will be needed. This will be made especially compelling because many of the foundational aspects of our lives here on Earth—things that we have learned to rely on to the point of taking them for granted—will be left behind forever.

Explorers on Earth, no matter how difficult the journey, can at least be comforted in being able to breathe fresh air, to experience the sights, sounds, and smells of nature, and even be reassured by the familiar tug of gravity. Extraplanetary settlers will have no such assurance of familiarity, and the resulting stress can have widespread negative consequences if not understood and controlled. This will be exacerbated by the fact that space journeys of today—and in the near future—are undertaken by small groups of select and very highly trained professional astronauts. High standards of motivation, discipline, dedication to duty, and expertise are accepted facts. Larger groups of people that will be needed for journeys of colonization will almost certainly exhibit a much wider range of variability in all these traits. Such diversity can be beneficial in many ways, but in the initial flights of these more ambitious undertakings it is not clear how to balance such diversity with the known successful approach to assembling an astronaut team.

Let us begin our journey out to the planets of the future with a look at current thinking for astronaut well-being followed by some informed conjectures as to where that might lead us.

SPACEFLIGHT HAZARDS TO HUMAN WELL-BEING

To go about systematically addressing the major risks to human health and performance in long-duration spaceflight, it is useful first to delineate these risks. The approach that NASA currently takes is to identify the main spaceflight hazards (environmental conditions) and then determine the specific human risks associated with each hazard (Francisco and Romero 2016).

1. Hazard: Altered gravity level (in space or on a planet other than Earth)
    a. Spaceflight-induced intracranial hypertension/vision alterations
    b. Renal stone formation
    c. Impaired control of spacecraft/associated systems and decreased mobility due to vestibular/sensorimotor alterations
    d. Bone fracture due to spaceflight-induced changes
    e. Impaired performance due to reduced muscle mass, strength, and endurance
    f. Reduced physical performance capabilities due to reduced aerobic capacity
    g. Adverse health effects due to host-microorganism interactions
2. Hazard: Hostile and closed environment
    a. Acute and chronic carbon dioxide exposure
    b. Performance decrement and crew illness due to inadequate food/nutrition
    c. Injury from dynamic loads
    d. Injury and compromised performance due to EVA operations

e. Adverse health and performance effects of celestial dust exposure

f. Adverse health event due to altered immune response

g. Reduced crew performance due to hypobaric hypoxia

h. Performance decrements and adverse health outcomes resulting from sleep loss, circadian desynchronization, and work overload

i. Reduced crew performance due to inadequate human-system interaction design

j. Decompression sickness

3. Hazard: Isolation and confinement

a. Adverse cognitive or behavioral conditions and psychiatric disorders

b. Performance and behavioral health decrements due to inadequate cooperation, coordination, communication, and psychosocial adaptation within a team

4. Hazard: Distance from Earth

a. Adverse health outcomes and decrements in performance due to in-flight medical conditions

b. Ineffective or toxic medications due to long-term storage

5. Hazard: Deep-space radiation

a. Risk of space radiation exposure on human health

THE RISKS TO HUMANS

It is easy to see from this list that almost every system in the body is potentially impacted by long-duration spaceflight. Note that these risks also include psychological and interpersonal issues that might arise due to the confines of any practical spacecraft for the foreseeable future.

NASA and its Human Research Program work to mitigate the major risks shown in the list. Some are more critical than others, and their criticality and priority depend on the type of mission. For example, the likelihood of a medical problem or a teamwork issue will be higher with a Mars mission simply due to the longer mission duration and the need to deal with in-flight problems without help from the ground. In fact, the majority of the problems are worse with longer and farther missions such as one to Mars. On the other hand, an issue such as sleep impairment might be *less* critical on a longer mission. Sleep is disrupted on the ISS because of a high workload and operational pace, occasional emergency procedures, and altered light-dark cycles. On a deep-space mission lasting months or years, it might be that a normal operational pace could be achieved during the journey itself, enabling a more normal sleep pattern. This is critical because of the key role that sleep plays in so many other systems, not to mention its effects on performance.

The most ambitious mission scenario now in the planning stages is a three-year mission to Mars, which would entail up to eighteen months on or near the planet. The most important risks in that mission fall out easily from the list above, given the duration, distance, and high degree of crew autonomy. This issue of crew autonomy is an especially important one to which we will return later. The key risks for Mars are the following:

Radiation. Galactic cosmic rays and solar particle events are the two main categories of radiation in deep space, and their levels are higher there than on Earth or in low-Earth orbit. There is a long-term increase in the lifetime risk of acquiring cancer from this radiation. This is not a major *operational* concern for a mission that might last just a few years, but it can lead to significant burdens on healthcare systems in longer-duration flights of many years unless proper shielding is in place or other countermeasures implemented. Of more immediate concern during a mission itself are its degenerative and cognitive effects, which are not yet fully characterized, but may occur with chronic exposure at lower radiation levels than the known cancer risks (Parihar et al. 2016). Nevertheless, the prospect of a crew being exposed to a solar event that induces radiation sickness and immediate cognitive decline is sobering. This is especially true when it is recognized that a deep-space crew must have a high level of autonomy due to remoteness from Earth. The radiation-related risks are a major concern with both short-term and long-term consequences.

Cognitive and behavioral issues. Isolation and confinement for long periods of time with a small group of people are challenging even for small, highly trained and dedicated crews. Obvious problems include disagreements with other crew members. But potentially more dangerous are difficulties with teamwork that can arise from these interpersonal issues that can have a direct negative effect on the success of the mission and the survival of the crew. This is an issue of particular import because team-work problems can be exacerbated by cultural differences and misunderstandings about individual roles on the crew. Given the desirability of personnel diversity, these will undoubtedly be major factors in future expeditionary and colonization crews. Related to these are effects on cognitive function, which is negatively impacted on long flights for reasons that are not fully understood. This is sometimes referred to as "space fog" and it is a general perceived slowing of mental processes. Space fog may be related to high workload, elevated CO_2 level, altered sleep cycles, and other stressors inherent to current spaceflights. Awareness of this issue is a key aspect of minimizing its impact, and professional astronauts are so highly screened that even with a decrease in cognitive function they are still high achievers. Nevertheless, the prospect of a crew undertaking a demanding mission at less than full cognitive capability must be considered.

Gravity (or the lack thereof). Due to the lack of gravity (more properly, the lack of a net gravity or inertial force vector), astro-nauts experience a shift of body fluids toward the head. This has been recognized for decades, and results in sinus congestion, puffy faces, and spindly legs, among other more serious issues. One of these is the relatively recent finding of changes in visual acuity after extended stays on the ISS. This is, as current thinking goes, due to the small but constant elevation of fluid pressure inside the head, which finds its relief by moving fluid down the optic nerve toward the back of the eye. This pushes on the back of the eye and distorts its shape, changing its optical properties. This is serious enough when dealing with high-performing individuals in demanding situations, but the prospect that it might be just an early indicator of lasting neural damage that accrues over longer periods of time in space is especially troubling.

Altered (reduced) gravity loading on the body also leads to a host of other problems. Constantly fighting against gravity while on Earth, so as to maintain upright posture and sufficient blood flow to the brain, provides a natural continuous form of exer-cise. This helps to maintain muscle strength, bone integrity, and cardiovascular function. In space, these physiological functions deteriorate unless measures are taken to protect them. Among the most effective measures is exercise, as discussed below. With proper exercise, astronauts can now return to Earth with sufficient muscle tone, cardiac function, and bone density. There is a caveat, however: While bone-mineral density is currently well preserved, it is not clear that bone strength (fracture resistance) is preserved, since bone is a complicated entity whose strength depends on its internal structure, not just its density.

Food. The ability to provide nutritious and enjoyable food is also a concern. The shelf life of foods is limited by storage technology, processing, and inherent chemical processes. The food problem is more, however, than simply making sure that caloric content and balanced nutrition are achieved, particularly since packaging for preservation and storage often sacrifices palatability. Food is one of the very few pleasures that crews can enjoy during flights. It is a familiar and comforting reminder of home and the strong social ties attached to group dining are important for crew cohesion. Thus, a variety of healthy and pleasurable foods must be provided or grown in the spacecraft or habitat.

Medicine. The likely occurrence of a significant medical condition is another major concern. The crew will consist of a small number of people who are highly fit, maintain healthy lifestyles, and are provided with opportunities for exercise and proper nutrition. Nevertheless, over the course of a three-year mission during which they will be exposed to a dangerous environment and called upon to be very active in new and unusual tasks, it is likely that medical concerns will develop. These might be as conventional as abrasions, sprains, and related simple injuries, or as complex as systemic changes initiated by fluctuations in the gut microbiome due to altered gravity and diet, and medication use. Head and back pain is common even on shorter flights to the ISS due to headward fluid shift and decompression of the spine. Rashes and minor injuries from floating debris are common. Particularly troubling is the increased possibility of kidney stones, either in space or on reaching a planetary setting, since the calcium lost from bones is excreted in urine and may form stones. These are just some of the issues that are known, as are many others which so far have been treatable and have not required a medical evacuation from the ISS. Simply due to increased time in mission and the challenges of going to Mars, however, it is likely that medical issues will arise and so a systematic and realistic approach must be taken. As one example of the tradeoffs involved, consider in-flight surgery. Current NASA thinking is that the amount of supplies and apparatus needed to perform even minimal surgery in space—not to mention problems of maintaining a sterile field and controlling blood and other fluids—will preclude surgery for the foreseeable future. This may change as noninvasive procedures are refined and the feasibility of surgery in resource-limited settings increases, but a realistic appraisal demands that consideration be given to the possibility that one or more crew members will not survive an entire mission because of the limitations of practical medical care.

A comparison might be made to other cases of medicine in austere environments, such as Antarctic bases and nuclear submarines. This comparison can be instructive, but there are limitations. In extreme cases, medical evacuations can take place from each of those settings, although this is quite challenging for Antarctica. The diagnostic equipment and medical supplies available on each site, while limited, are more plentiful than on a spacecraft. Perhaps of most concern, communication delay due to distance from Earth will preclude most forms of medical support from the ground, especially in an emergency. Thus, the crew will need to have the onboard resources to not only treat medical conditions, but also to diagnose and anticipate them, and to make complicated clinical decisions based on the latest information without contact with Earth.

Information. Finally, crews will need access to a great amount and variety of information while on their journey: personnel, vehicle, and mission status, for example. Many complicated systems will need to be monitored and maintenance performed in a timely fashion. For all of this, an intuitive set of human-system interfaces will be needed. These will have to provide crews with necessary information that is easy to digest in a timely manner and that helps guide them to proper actions. Too much or too little information is not helpful and can lead to burnout or a dangerous lack of situational awareness. Anyone who has dealt with current information technology and tried to make sense of the seemingly haphazard information in an internet search, which is easy to obtain but difficult to process and understand, will know that this area still requires significant work.

There remain other considerations, but this brief survey should give an idea of the range of concerns for humans on deep-space missions that are in the current planning stages. Countermeasures are in place or in development for each of these issues. We note just two here. Exercise is a very powerful and successful countermeasure and is in regular use on the ISS (approximately two hours per day per person). Not only are there obvious physiological benefits for bone, muscle, and cardiovascular fitness, but exercise provides a sense of well-being as a psychological countermeasure. "Runner's high" is well known, but also the exercise schedule provides "protected time" during which other tasks are secondary, allowing a mental break. (Astronaut time is the single most valuable quantity during a spaceflight, and their time is often scheduled to the minute. This can lead to timeline pressure and constant stress.) Nutrition is the other main countermeasure now in place and it too has wide-ranging benefits. Again, not only are there obvious physiological benefits, but food provides something familiar and comforting in an environment that is largely devoid of the normal human pleasures. It is also a socializing aspect, which crews find to be very important when the demanding schedules on the ISS can inhibit regular social interactions.

Note that in the area of psychological function, including teamwork, the countermeasures are not so highly developed. Education and awareness are perhaps the most useful current approaches. It is worth noting that one very important psychological countermeasure, however, will not be available to crews on Mars missions: the ability to speak to someone on Earth at any time about any issue. This includes family and friends, but also physicians and psychologists. The communication delay on a Mars mission will preclude most meaningful conversations of this nature.

(More information on these risks and countermeasures can be obtained from the series of Evidence Reports that the NASA Human Research Program maintains on its web site: https://humanresearchroadmap.nasa.gov/evidence/)

AUTONOMY

Thus, we come to the issue of crew autonomy. This is not a major concern with missions to low-Earth orbit, where home is just a few hours away and there is constant communication with mission control. In an emergency, the crew knows it can get instant and expert assistance. For missions farther out into deep space, autonomy will be a critical issue for Mars and even more so for missions beyond that. The need for autonomy increases with distance and duration to the point where it will be almost beyond our current comprehension for missions in the next century that might settle and colonize other planets.

The main feature of autonomy is that the crew must be able to function on its own under nominal and emergency operations. The ability to recover from the unexpected and maintain mission success is closely related to resilience. It would be nice, then, if resilience could be characterized and even measured, so that crews would know when it is being compromised. Let us consider how this might be done.

THE WORST-CASE SCENARIO

From the previous discussion, it would appear that things are under control for humans in missions to LEO and maybe even to the Moon and Mars. NASA's ambitious but feasible plans call for the mitigation of each major risk described above through extensive research and proper mission design before undertaking a lunar or Martian journey. Consider, however, the following scenario on the way to Mars:

There is a solar flare or coronal mass ejection—a solar particle event (SPE) that emits a storm of high-energy protons. There is an undetected microorganism hiding, deeply embedded, in the spacecraft water supply, or the food supply, or the air-filtration system. The spike in radiation induces a mutation in the microorganism. At the same time, for reasons not completely understood, the astronauts' immune systems are undergoing alterations that make them less effective in fending off some pathogens. Also, at the same time, one person is on a course of antibiotics that has caused a significant change in his gut microbiome, which modulates many body functions. This has seriously depleted the on-board medical stores. As if this weren't enough, the exercise equipment breaks. At first glance none of this is particularly troublesome because, as we know, each of the individual risks has been sufficiently mitigated: radiation, food supply, immune function, microbiome, medical supplies, and exercise countermeasures. So, there is no problem, right?

Wrong. Two days later the entire crew is dead or debilitated to such an extent that the mission is compromised. What went wrong? The radiation spike induced a mutation, which impacted the nutritional status of the crew (food or water supply). The human immune system is altered in space, as are some pathogens, with some becoming more virulent and others less so. If a mutated organism invades a compromised immune system, the outcome is not likely to be pleasant. The gut microbiome—the many microorganisms that live in the intestinal tract—exhibits some alterations when in space, and medications like antibiotics can also have a devastating (hopefully temporary) effect on the quantity and diversity of these microorganisms. The microbiome has been implicated in a wide variety of physiological functions, including even cognitive status. Exercise, in maintaining general systemic health, could help overcome the changes in immune, microbiome, and cognitive function. When these events transpire together (and they might be related to each other, as for example a radiation event might adversely affect instrumentation that runs the exercise device), unanticipated consequences can ensue.

Thus, a crazy interaction of factors has conspired to bring about a devastating outcome. But this is not as crazy as it sounds. Aviation and aerospace "mishaps" and incidents are rarely caused by a single precipitating factor; they are almost always the consequence of an interacting series of events. In our example, the interactions have not been understood, characterized, and tracked properly over the course of the mission. Early recognition of changes in these factors, and how they might be connected to each other, might have led to an early intervention to stave off the devastating result.

How can we do this? How can we systematically track all the of the relevant factors and their interactions and predict when the overall system (mission) is approaching an undesired state from which recovery might not be possible? How can we use this knowledge, in other words, to increase mission resilience?

A BETTER WAY

Imagine now a situation in which the on-board computer contains a mathematical model of the main factors that can be measured and tracked over the course of a mission: environmental and physiological parameters, interpersonal interactions, sleep quantity and quality, mood, exercise, and many others. Sensors for many of these quantities exist, as do "wearables" to make individual personal measurements continuously and nonintrusively; additional new sensors are being developed all the time. The mathematical model also contains information about expected interactions between parameters. For example, a change in CO_2 level can be expected to have some specific effects on the crew, and that corresponds to an interaction or link between atmosphere and psychology, mood, depression, and team cohesion, in addition to performance on standard tasks. If CO_2 level changes for some reason, the computer can compare the effects that actually occur to the expected effects. The actual effects might not be pleasant or desirable, but if they are at least understood, then the mission is still on track: The actual mission falls within the range of the model. If there are unexpected interactions, such as an unusual physiological response on the part of some of the crew to a change in CO_2 level, then this can indicate that a crucial interaction has changed, and the underlying model of interactions is no longer accurate. If this happens with enough interactions, a problem might be in the offing. This is because, recall, it is unanticipated and little-understood interactions that lead to bad outcomes. (The model and algorithm described here are beyond the current state of the art. They might be based on complexity theory and network concepts, with AI and machine-learning components, some of which have yet to be developed.)

With this sensor-model system in place on a spacecraft to assist the crew, now consider the following scenario. The level of CO_2 goes up for reasons that can't be helped (perhaps the scrubber is down and there is a need to conserve consumables and power). As expected from elevated CO_2, two crewmembers get into arguments and have difficulty working together, but one other crewmember exhibits no change in behavior when she typically becomes irritable and sleeps longer. This is a change in interactions—in cause and effect. The cabin CO_2 might still be within normal limits, and the individual crew behaviors might be within normal acceptable ranges. The interactions, however, tell a different and subtler story. Is the third person already maxed out with other stresses such that elevated CO_2 no longer has much effect? Is there some other change in metabolism that has altered the response to CO_2? Whatever the cause, this altered interaction should be tracked, and if similar unusual changes occur soon after, then some type of intervention might be needed to fend off an undesired developing situation. Perhaps the computer algorithm recommends a change in the schedule to separate the two who are fighting, and additional rest and relaxation for the third, and then monitors the situation to see if these changes improve crew conditions and interactions.

Resilience and performance have now been enhanced. We have provided this crew the tools to be resilient and they have maintained mission resilience independent of mission control. This is a microcosm of what will be necessary on a much larger scale for the much more ambitious journeys of colonization.

A NEW BREED OF COUNTERMEASURES

Considerable thought and effort have gone into defining and developing countermeasures, as noted previously. There is a rigorous—if sometimes contentious—process for determining what areas need to be addressed and what the best countermeasures are. We can learn from this process in extrapolating to future missions what the major problems might be and how they might be mitigated.

Consider a journey to an outer planet (or its moon), or even to another solar system, possibly taking several decades.

If there is no artificial gravity (AG), then the physiological degradations can be substantial and dangerous—a serious concern once the crew reaches its destination. Given current knowledge and a set of reasonable conjectures on the design of such a mission (crew makeup, spacecraft capabilities), it is likely that crews would land in such a debilitated state that their very survival would be jeopardized. Consider first that on the ISS exercise is a major countermeasure for an array of deficits, but in current implementations requires approximately two hours a day. Let's allow that sophisticated methods are developed to augment the potency of exercise, such as ketogenic diets, blood restriction, and muscle cooling. Even with these in place to reduce the actual exercise requirement, it is likely that compliance will be an issue. Further considering the psychological difficulties of such a journey, motivation will flag, and the drive to maintain health could in turn diminish.

One has only to observe astronaut crews returning from ISS missions of six months to see the possible consequences of the problem. Even in these groups of elite, highly trained and motivated professionals, with excellent countermeasure compliance, there is need for assistance for the crew in leaving the spacecraft and making its way to the medical tent. The ability to walk in any meaningful way is impaired for hours to days, and full recovery can take weeks or months. This is the current best-case scenario. A possible solution to this is to allow the first planetary settlers a significant period of rest, recovery, and rehabilitation upon landing. The spacecraft would essentially become a rehab facility for physical therapy while the occupants once again learn to walk and maneuver in a gravity field and regain muscle strength and aerobic capacity. This is a reasonable approach but requires a much larger habitat on the surface than would be needed simply for landing a group of people.

Even in this optimistic view, there are physiological issues such as loss of bone density and strength that will remain significant and dangerous. (Landing on a body with less than a one-gee field will ease these concerns since the likelihood of a fall, and the intensity of a fall, would be reduced and hence the possibility of breaking a bone would be reduced.) Added to this is the chronic shift of fluids to the upper body and head, which might not only induce vision problems as seen in some ISS astronauts, but could be causing low-level neural damage. The prospect of a large number of slightly demented individuals stumbling around on a new planet, tripping and breaking bones, is not one that most planners would find desirable or acceptable. They would be like the first settlers in a foreign land who encounter a bacterium or virus for which they have no natural immunity, only in this case the scythe is wielded by gravity and not by an organism.

Artificial gravity is therefore a compelling option for such a journey (Young 1999; Clément and Bukley 2007). The longer the journey, the more compelling the case. When we consider journeys of decades or more, it is almost unconscionable not to consider AG as a requirement. Indeed, many of the practical issues that inhibit its ready acceptance in current planning should have been solved by the time these journeys of colonization become a reality. But AG is not a panacea, nor is it obvious how best to implement it. One appealing method entails constant linear acceleration of the spacecraft for half of the journey, and then constant deceleration (technically, acceleration in the other direction) for the other half of the journey. This creates a linear acceleration force, which is indistinguishable from gravity itself. The change from acceleration to deceleration at the halfway point would be quite an event. Other than that, this would be a benign method to produce some level of AG in the entire craft. The propulsion requirements for this are currently prohibitive. Better to think along more conventional lines: a rotating craft that produces centripetal force proportional to radius (distance from the axis of rotation) and to the square of the rate of rotation. Larger and faster spinning produces more AG.

Rotation of an entire spacecraft in this manner is daunting. It also raises the question of what happens when the craft reaches its destination and rotation stops or the crew departs the rotating craft and experiences zero-gee for the first time, at exactly the time they must prepare for a challenging planetary descent. Perhaps there is a better way. Much thought has been given to alternate approaches, such as rotating just a part of the spacecraft. If the sleeping quarters are rotated, the crew will experience eight hours of AG each night, while enjoying the pleasures of zero-gee the rest of the time. A spacecraft might also contain a small exercise chamber in the form of a cycle ergometer (exercise bicycle) that moves in a tight head-over-heels circle when pedaled, supplying exercise and providing AG at the same time. Each of these approaches has its problems, but some combination of them could almost certainly be made available to a large crew.

Even with some combination of these implementations, however, some problems remain. We don't yet know how much AG—what level and for how long—is needed to counteract the deconditioning effects. With research this can be determined along with the engineering solutions. The benefits would appear to be significant. AG would be almost ideal for bone and muscle effects, and most especially for the problems attendant to headward fluid shifts, which cannot be countered by exercise. For some other areas, there are unfortunately undesired side effects. The vestibular system, for example—the balance organs in the inner ear—can adapt well to a zero-gee environment but is used to orient within inertial reference frames. *In other words, rotating environments are not kind to the balance system.* Anyone who has spun around a few times on a bar stool (sober or otherwise) and then tried to stand and walk understands the problem viscerally. Several playground devices have similar effects. Humans can adapt to these challenging settings (Lackner and DiZio 2003), but the transition to and from an AG situation could be difficult.

But wait—are we missing something here? Is it possible that there are salutary effects of zero-gee itself—effects that would be missing if AG were implemented? Consider that spaceflight itself and the spacecraft environment with its close quarters, strenuous workload, and imminent danger are demanding enough in and of themselves. Is it possible—just possible—that the experience of weightlessness in this setting provides one of the few pleasures that helps make it tolerable? Would the imposition of AG in such an already stressful situation actually make things worse? We do not know, and the idea might be far-fetched, but it is worth considering. There are pleasurable effects from unusual patterns of vestibular stimulation—witness the popularity of roller coasters and other semi-nauseating escapades. Likewise, the experience of short periods of zero-gee in parabolic flight has been described as the most fun one can have in public—at least short of actual spaceflight. The Apollo astronauts remarked that they could each find their little bit of personal space in the cramped spacecraft, since they were free to explore all three dimensions in zero-gee and could inhabit nooks that were otherwise inaccessible. How much of the famous "overview effect" is due to seeing Earth from above, how much from the experience of $0g$, and how much from their interaction? (The overview effect is the vivid realization reported by many astronauts of the fragility and isolation of the Earth in space, the invisibility of political boundaries, and the sense of union among the people on the planet (White 1998).) These are open questions and although it might be small it is too soon to dismiss the possible beneficial effects of zero-gee itself on psychological well-being. Without it, a spaceflight journey of decades might become mere drudgery rather than an adventure— one may need a reminder that the environment, and hence the journey, is unique, and the constant reminder of zero-gee could do that very effectively (despite the problems that it also creates).

Let us return to the more practical aspects of artificial gravity. As noted, AG can address in a natural manner many of the physiological concerns of extended spaceflight. However, even with AG there will be problems. Currently the biggest risks identified by NASA are related to radiation exposure and adverse effects of spaceflight stressors on cognitive function and behavior. AG will not help these. Radiation might be addressed with clever shielding and maybe even radio-protectant pharmaceuticals. But make no mistake: Psychological maladjustment is the potential killer and this problem will not be solved with AG.

PSYCHO KILLER

Consider the psychological issues arising during a decades-long journey (Manzey 2004). Some will be inherently mitigated relative to even the first Mars missions, because of a larger spacecraft with more people on board. Thus, the problems related to confinement might be lessened, although it's still likely that the minimum tolerable amount of volume per person will be implemented—it is just that that volume will be used to better effect: It will be reconfigurable for the changing needs of the crew, allow group interactions or individual solitude as needed, and provide for some individual psychological needs. The presence of many people can reduce the problem of small groups becoming tired of each other and getting on each other's nerves; of course, other problems are raised through the greater diversity of skills, motivations, and dedication. Teamwork might be aided by the ability to rearrange teams for some tasks through cross-training of individuals so that the same small group does not have to work together all the time.

However, remoteness and isolation will continue to be significant issues. There were dire predictions in the early days of human spaceflight, to the effect that astronauts might experience a dramatic detachment or breakaway phenomenon, being physically separated from Earth to an extent never before experienced—beyond the atmosphere. This concern turned out to be misplaced, even overblown. Instead, it turned out approximately six decades later that astronauts take great pleasure in viewing Earth from the cupola of the International Space Station. Many have commented over the years, even before the ISS, that watching Earth was a great treat, possibly even therapeutic (not in so many words). This consoling aspect will not be available to those on long-distance missions. Even in going to Mars, the image presented of Earth will be that of just another star in the heavens. The psychological impact of this is impossible to determine. Add to that the realization that there is no return if the journey is one of long-term exploration or colonization, and some form of separation anxiety might well occur. There will be the need to change one's loyalties and sense of identity to a wholly new place.

Until this mental transition fully takes place, there may be some erosion of resilience—a relative deficit in performance and in dealing with problems—at least in the middle term of the voyage as memories and longing for Earth and what it represents remain strong. Certainly, in a generation or two this reminiscence will remain but perhaps it can be modified and infused with feelings of attachment for the new celestial home. This is not a "mere" psychological factor—it has an impact on several factors which, as we have seen, are tightly connected. Thus, in the mid-term, it may be necessary to have Earthlike reminders that gradually shift in a weaning process. It has been proposed, for example, that deep-space missions now on the horizon include a virtual-reality arrangement that mimics the traveler's home—his or her favorite room, settings, people, plants, animals, and so on. One cannot help wondering if this is wise. Would not a "clean break" be better? For early missions there might be no alternative, but later it might be better to break this connection with Earth as soon as possible. (NASA and other space agencies use a variety of Earth-based analog facilities in an attempt to mimic and mitigate these psychological concerns. They all, however, fall short in some way from reproducing the full range of effects: The missions are not long enough, the setting is not inherently dangerous, and the facility is not truly remote, or it is so large that the crew is not confined in the same sense as in a spacecraft. In this critical issue we will not truly know until we go.)

How will we track if these measures are effective, even the psychological measures? These aspects have implications for physiology and performance through stress pathways. So, just as proposed above for shorter missions, the continuous monitoring of physiology and performance, including psychology and inter-personal relations, will be one key to tracking the effectiveness of breaking one's loyalties from one planet to a new one.

INTERACTIONS FOR THE FAR FUTURE

The psychological issues are bad enough. But all this discussion does not even touch the real issue. The likely worst case is that there will be a confluence of circumstances that is unforeseen: a set of events in which several risks come under attack, each of which was considered to be adequately mitigated. Examples have been given previously in the context of a mission to Mars, but this will take on increased importance in the case of longer and farther missions in which the crew will be on its own. Reliance on Earth might be realistic for the first few weeks, after which it will be a rueful wish and then nothing but a quaint memory. Crews and colonists will need the ability to maintain resilience—mission success in the face of unknown perturbations—apart from the assistance of anyone on Earth. And as we have seen, this will be in large part a consequence of paying due attention to interactions among factors (Shelhamer 2016; Mindock et al. 2017).

This takes on a whole new meaning in the context of colonization, when the travelers will truly separate themselves from Earth and its support mechanisms (Bell and Morris 2009).

Faced with the prospect of never returning to Earth, psychological issues will loom large, especially on the first of such missions. As noted, this has widespread effects since stress is a major factor with many connections. There will also be physiological and psychological changes that occur once ensconced on another planet or moon. These might impair the ability to return to Earth, which would be another reminder that that is no longer a viable option. The question then becomes whether to allow these adaptive alterations to proceed or to attempt to slow or inhibit them. The answer is not a simple one. Adaptive evolutionary changes take place on Earth over long timescales, partly because the environment is relatively stable (certainly as far as gravity is concerned). Faced with a dramatically different environment—altered gravity level, unfamiliar atmospheric pressure and composition, different magnetic field, to name a few—evolutionary processes in the human organism might be accelerated. Under such circumstances, epigenetic alterations might take on a larger role in the heritability of acquired traits. Whatever the mechanism, settlers will likely be faced with the problems inherent in rapid change—only this will involve changes to the humans themselves. The possibility that some of these changes will be undesired—and could interact with other changes to the overall detriment of the person—should not be ignored. To the extent that these adaptive alterations proceed, some monitoring might be in order so that undesired and unanticipated interactions can be identified.

It is almost certain that some adaptive changes will prevent the person from ever returning to Earth, or even to a planet with a different gravity. Consider landing, settling, and evolving on a planet with considerably less than the one-gee of Earth: This might eventually lead to taller and thinner humans, since maintenance of blood flow to the head would not be as challenging. Cardiac capacity would change for the same reason. The long, weight-bearing bones would also become thinner and weaker, as would the supporting musculature. In short, organisms are good at shedding unnecessary metabolic costs to make efficient use of resources such as nutrients, and these are all appropriately adaptive changes for the environment in question. But these people would then not be able to function normally on Earth, where they would be highly prone to injuries. It would be unrealistic and unethical to even consider sending them (or their offspring) to Earth or other locations where the gravity level is significantly higher. Thus, humans might indeed become a multi-planetary species in this way, but each subpopulation of humans would be forever tied to just one or a small subset of planetary bodies: No single individual would be truly multi-planetary.

But also consider less dramatic effects. Taller people could be problematic in habitats designed with small dimensions to preserve resources. It would be well to track and predict these types of interactions between physiology and engineering.

On the other end of the spectrum, there might be a temptation to enhance or accelerate adaptations to specific environments. Or to simply make immediate alterations for expediency. This raises the specter of genetic modifications, or surgical ones. One might argue that many space settings would call for the shortening or removal of the legs. In a constant zero-gee setting such as a permanently orbiting station, legs are unnecessary for locomotion. In fact, they can be a hindrance by banging into things, especially as proprioceptive sense of leg position decays from lack of use. The resulting reduction in body mass and fluid reservoir would be beneficial in reducing radiation exposure and in combatting headward fluid shift. On the other extreme, on a planet with a very high gravity, metabolic costs could be diminished by reducing the hydrostatic gradient—that is, by making people significantly shorter.

As intriguing and tempting as these concepts might be, society best tread lightly in any such endeavor at the risk of introducing new complications or overlooking crucial interactions. The introduction of new species to an isolated environmental niche is a historical example: Where there is no natural predator the new species overtakes the available resources. This is one simple form of unintended consequence. There are many others, and in a space-settler setting where there is precious little backup capability (you can't go home again), even subtle second-order effects can take on outsized significance.

CONCLUSION

Thus, the key in all these situations of long journeys of settlement or colonization is to recognize that we might—just might—be smart enough to mitigate the major known risks for long-duration spaceflight on an individual basis for a relatively short Mars exploration mission (three years), but we are unlikely to be smart enough to determine in advance the countermeasures that will be needed on journeys of colonization. It is almost a certainty that unexpected and unanticipated problems will arise—perhaps a new form of psychological syndrome caused by an unusually strong attachment to an artificial habitat that takes on undue importance in the absence of a familiar Earth and a viable atmosphere, displacing emotions and bonding with other humans. This is pure conjecture of course. More likely, perhaps, would be a novel combination of radiation that interacts with an organism in the soil and revives a dormant species (we see viral shedding on the ISS) for which the weakened immune system is no match, while at the same time the medical supplies have been depleted in treating more conventional problems.

So, how do we give crews the tools to deal with these larger problems—the unknown unknowns? What would these tools be? Some are tangible and, while not trivial, are at least easy to delineate in principle: 3-D printers, DNA sequencers, medical instrumentation and diagnostic equipment, a vast database of information and the ability to acquire updates (not a trivial matter when communication with Earth is challenged by time lag), and information systems that provide the crew useful and important information in a timely manner without saturating them.

But more fundamentally we need to provide the *conceptual* tools for dealing with the unknown. These are the same as described previously but at a higher level of complexity. This essentially entails a mathematical model that encompasses a deep understanding of the many factors that impact survival and mission success. These are, as noted, not only medical, physiological, and psychological, but also encompass interpersonal interactions, habitat configuration, task planning and design, scheduling, and many others. Sensors for these key variables can track the most important of these factors, continuously feeding data to the model, which would monitor each individual parameter for problematic deviations but also track interactions between parameters and compare them to what is expected from its stored database. Adding to this complexity is the fact that this model must adapt as the people adapt to their new setting and understand when significant changes are part of a beneficial adaptation process versus a detrimental maladaptation or dysfunction.

Thus, we must provide crews of the future the *tools* for solving problems and not the *answers* to the problems *per se*. We can teach a crew to fish and it may survive for a year, but if we give the crew the tools to make rods and reels and find fish and adapt in order to metabolize other types of fish, then they can survive and thrive for generations.

EPILOGUE
RISK AND RESILIENCE AS A SPACEFARING SOCIETY

Many of the issues raised here transcend those of engineering and operations, the typical realm of advanced spaceflight discussions. They become issues that the larger society should address. As a society, are we willing to do what it takes to enable these voyages of colonization and settlement? The concern is not just the financial cost or the opportunity cost, but the larger cost to society in terms of resource allocation and even more so in terms of *perspective*. Being audacious enough to give people a fighting chance of surviving and thriving on other planets might mean, as indicated here, that changes in human psychology, and perhaps even physiology, might be needed. New structures of governance and civic cohesion might also be needed. All these changes might in fact occur on their own as a natural adaptive response to a new environment, whether we like it or not. Yet we retain the right to decide whether to put our fellow humans into that situation. How might the attendant physiological and psychological changes in turn change our view of what it means to be human, when it no longer explicitly means "Earth-dweller"? Will we recognize governing structures that are designed to accommodate small populations and environmental stressors on an alien world, and will we be comfortable with them as representative of the civic decisions that have guided our institutions on Earth? Are we ready for such a change in how we see ourselves?

It is one thing for government space agencies, or private companies like Blue Origin and SpaceX, or futurists, to ponder these issues—even to make pronouncements as to preferred policies (as we do in this volume). However, if humans are truly to colonize and settle elsewhere in the solar system as a species and not just as a small group of rugged individualists, then society must ponder these issues in an open forum and attempt to reach some consensus. To not do this will leave the decisions to those organizations—public or private—that first have the means to undertake the journeys.

Consider a microcosm of this larger issue. Spaceflight is a risky endeavor. It often results in the loss of life, and there is a general acceptance of this fact as a society. We have decided to accept that risk. It is important to recognize that this is not just a set of individual choices made by individual astronauts (Kahn et al. 2014). Astronauts might, for example, be willing to accept a large potential increase in lifetime cancer likelihood in exchange for trips into deep space. As a society, on the other hand, we might not accept this risk (through the decisions made by the space agency as dictated by law and regulation).

If people want to risk their lives on spectacular feats such as space travel, do we, as a society, have a right to stop them? Do we have a right, or even an obligation, to withhold resources that might aid in their success? Does it matter if the people involved are normal citizens like us, and not those specifically trained to recognize and evaluate the risks? What makes an astronaut remarkable and commendable is the willingness to undertake a great risk while fully understanding the nature of the risk; that is part of their job and we respect them not for being reckless but for being aware.

These are not moot questions. Society benefits from great feats successfully accomplished. Society suffers the consequences of a spectacular and fatal failure—especially if it impacts future policies and hinders further progress through risk-aversion. Even beyond this, if the decision to support these endeavors is made as a society, then a great number and variety of societal institutions can marshal themselves to the cause: educational, civic, research, and financial. In this way, the resilience of the endeavor is enhanced through multiple institutions and societal structures contemplating and developing multiple simultaneous solutions and approaches (rather than a more constrained and narrow approach that would be feasible under the auspices of a small group without that broader societal support).

Fortunately, we are in a position to take incremental steps in this direction, and they are taking place now. Commercial suborbital space flight will be a reality soon. Passengers will be able to pay for a suborbital rocket hop that takes them into space (above about one hundred kilometers) for a few minutes of weightlessness. These trips will, especially at first, entail significant risk, regardless of the great amount of engineering and due diligence now being invested in making them safe. The decision in the United States, at least for now, appears to be that the less oversight and regulation for this new industry, the better. The concern, once the spacecraft have been tested and the passengers informed as to risk, is the safety of those on the ground who have not elected to be participants. This is an approach that places great responsibility on the flight operators and their customers, and implicitly asks society to accept the attendant risks. It remains to be seen if this approach holds up in the face of disasters, near-misses, or mishaps that are inevitable in this new venture. The decisions made at those times will tell us much about whether we are ready to face the greater challenges of planetary colonization.

REFERENCES

Bell, Sherry, and Langdon Morris, ed. 2009. *Living in Space: Cultural and Social Dynamics, Opportunities, and Challenges in Permanent Space Habitats.* Aerospace Technology Working Group.

Clément Gilles, and Angeli Bukley, ed. 2007. "Artificial gravity." Springer Science+Business Media.

Francisco Dave, and Elkin Romero. 2016. "NASA's Human System Assessment Process." Presentation at Human Research Program Investigators' Workshop (8–11 Feb). Galveston, TX.

Kahn, J. P., C. T. Liverman, et al., ed. 2014. *Health Standards for Long Duration and Exploration Spaceflight: Ethics Principles, Responsibilities, and Decision Framework.* National Academies Press.

Lackner, JR, Paul DiZio. 2003. "Adaptation to rotating artificial gravity environments." *Journal of Vestibular Research* 13, no. 4–6 (February):321–30.

Manzey, Dietrich. 2004. "Human missions to Mars: new psychological challenges and research issues." *Acta Astronautica* 55, no. 3-9 (August–November):781–90. https://doi.org/10.1016/j.actaastro.2004.05.013.

Mindock, Jennifer, Sarah Lumpkins, et al. 2017. "Integrating spaceflight human system risk research." *Acta Astronautica* 139 (October):306–12. https://doi.org/10.1016/j.actaastro.2017.07.017.

Parihar, Vipan K, Barrett D. Allen BD, et al. 2016. "Cosmic radiation exposure and persistent cognitive dysfunction." *Scientific Reports* 6 (October):34774. https://www.nature.com/articles/srep34774.

Shelhamer, Mark. 2016. "A call for research to assess and promote functional resilience in astronaut crews." *Journal of Applied Physiology* 120, no. 4 (February):471–2. https://doi.org/10.1152/japplphysiol.00717.2015.

White, Frank. 1998. *The Overview Effect: Space Exploration and Human Evolution.* AIAA.

Young, LR. 1999. "Artificial gravity considerations for a Mars exploration mission." *Annals of the New York Academy of Sciences* 871 (May):367–78.

At the Bottom of the White

Todd McCaffrey

New York Times bestselling author Todd Johnson McCaffrey wrote his first science fiction story when he was twelve. He has published over a dozen novels, including eight novels written in The Dragonriders of Pern universe and nearly twenty shorter works. His latest book, *The Jupiter Game*, tells of an alien first contact. You can learn more about him at www.toddmccaffrey.org.

YOU ARE LEAVING EDEN.

The words floating in the air rustled as Cin approached them. The letter "Y" glistened at the top and seemed to be turning itself into some stylized butterfly—*bio or software?* Cin wondered, her lips twitching upwards as she walked through them.

Eden was the safest place on the ship: a combination of oasis and hydroponics garden on a massive scale.

Cin craned her neck over her shoulder for a moment as she confirmed that—once again—the letters had completely rearranged themselves to say:

YOU ARE ENTERING EDEN.

As she turned back and continued her exit, another set of floating letters asked: ENTERING cross-contact AREA. CLOTHING REQUIRED.

She moved out of the walkway and into the alcove that was the changing area. She went to the dorm's locker and pulled out her shipsuit through the simple expedient of reaching in and pulling out the first thing she found.

Valrise knew her: She would never get the wrong clothes. She could have reached her hand into another locker and she would still have received the right clothes—and a polite admonition as well as an unvocalized bio-check to be certain that she wasn't suffering from some mentally debilitating malady.

Valrise was an old ship; she had quirks that newer ships might not value but... *Valrise* was an old ship. *An old ship is one that works.*

Cin was surprised to feel a bulkier sweater among her things. She was going somewhere cold, she surmised, shrugging on the sweater after pulling on her single-piece shipsuit. The sweater smelled new—clearly something *Valrise* had cooked up for the occasion and had Cin's name stenciled over the left breast.

"And some makeup," *Valrise* said. "There are Customers in the ice rink."

Cin gave her a questioning look.

"They aren't used to shaved eyebrows," *Valrise* explained.

"Makeup?"

"Use the dark pencil to mark your eyebrows and the synthetics will do the rest."

Cin felt inside her locker and pulled out a slim case. There was a mirror inside and a thin stick with dark substance on it. Cin had used makeup before—a couple of years back—so she wasn't totally unfamiliar with the notion.

"Lip color?"

"Natural will do."

"Anything else?" Cin asked after using the liner.

"Hair," *Valrise* said. Cin grinned as she looked in the mirror once more and saw that hair had sprouted on the ridge of her eyebrows. "Pull the hood of the sweater up over your head."

Cin complied.

"The young ones are competing," *Valrise* told her. "The Customers are quite amazed at our gravity control."

"How far back are they?"

"A good hundred, hundred fifty years," the ship told her.

"Poor, proud, and paranoid?" Cin asked, repeating the mantra that the crew used when referring to earthers.

"Exactly," the ship replied. "Now, go!"

Cin was glad of the sweater when she entered the much cooler confines of the ice rink. She paused as one of the contestants took to the ice, gained speed, and darted toward the low mound in the center. Cin's newly applied eyebrows descended thoughtfully as she judged the positioning of the skater: She was deliberately off the central axis.

Cin's observations were confirmed when the skater shot up the mound, entered a skew loop and neatly landed upside down on the inverted rink, and continued in a curved loop "down" the mound and onto the flat ice on the "ceiling" of the rink. Those of the crew watching applauded mildly while the more obvious Customers looked on in slack-jawed wonder—antigravity was always one of the first technologies to be lost when colonizing new worlds.

Valrise and the other trader ships spent some of their time reestablishing contact—and trade—with lost colonies. It was short-term and not as profitable as hauling goods between advanced worlds—the advantage was in establishing long-term trade. *Valrise*, in her hundred years of existence, had built up quite an extensive trading circuit. There was always danger in establishing new trading partners; care was required and friction inevitable as hopelessly outdated planets met up with the latest technology... and culture.

Hence the "dog and pony show"—as *Valrise* called it. Cin had been so intrigued by the phrase that she'd looked it up: She couldn't understand why a sentient starship would communicate with references to various four-legged mammals but she'd long ago come to recognize that the ship had a quirky sense of humor.

The Customers had been brought out for the day by *Lewrys*, one of *Valrise*'s fast shuttles. *Valrise* herself was still a good two weeks away from her flyby of the planet, so *Lewrys*' high-speed run had been its own sales pitch. *Lewrys* had already made the same run earlier when *Valrise* had first returned to normal space to reopen negotiations and arrange the present festivities aboard *Valrise*.

A second skater started on the downside rink and made the leap to the upside rink and then the two of them entered into a very precise—and risky—*pas de deux* between the two gravity fields causing not just the Customers, but also crew, to gasp in delight.

The world of the new Customers was named Arwon. The inhabitants, at least some of them, were Arwonese. The Arwonese had just recently completed a war of integration—another affectation of lost worlds that Cin didn't entirely understand (it was bad for trade, so why?). *Valrise* had provided a news *précis* which had left Cin mostly bored and slightly informed—she got the impression that the losers were in hiding and an inkling of some dark means of repression being used. Messy.

Cin had learned a number of years ago that the Customers could be evaluated by how conservatively *Valrise* had the ship's crew dress: These Arwonese were clearly among the more strait-laced worlders—every crew member visible was fully clothed and wore eyebrows and head coverings or actual hair.

"You are Cin, the bouncer, are you?" a stranger's voice asked from beside her.

"I am," Cin said. She found herself looking down into a pair of intense dark eyes set in a face framed fetchingly by raven locks of hair. There was almost an impish look to the...woman...Cin guessed from the shape of her body. Cin called up her implant and provided it with a feed from her optic interface.

She is the Calmt Prime, Valrise informed her through the link. A quick overview of the governing hierarchy and the woman's place in it caused Cin to blink in surprise: She was talking to the third most powerful person from the planet.

"And what is it like, bouncing?" the Calmt asked her.

Cin closed her eyes for a moment, recalling memories. She smiled as she said, "It is like nothing ever experienced before."

"We have developed skydiving in recent years," the Calmt informed her.

"That is when one falls in the air until restrained by a high-drag device?" Cin asked, trying to frame the words appropriately.

She must not have gotten it quite right; the Calmt smiled before replying, "Something like that."

Cin nodded. "Bouncing is more involved than that. Our bouncers are purpose-made to resist high temperatures while giving their occupants complete control over their position and angle of entry."

"And you bounce off the planet's atmosphere?" the Calmt asked, clearly trying to understand the allure of such an undertaking.

"Yes," Cin replied. "We have sensors to feed us and our bouncers are transparent. So we get to see it in real time."

"What do you see?"

"At first, nothing," Cin replied. "But as we get deeper and deeper into the atmosphere, the air around us begins to glow a deep pink with the heat of our transit."

"That must be quite frightening."

"Oh, no!" Cin said, shaking her head. "It is quite beautiful."

"Aren't you in danger of burning up?"

"We control our angle and our thermal bleed," Cin said. "The bouncers radiate excess heat away every time they leave the atmosphere."

"But it *is* dangerous."

Cin nodded. "Although I imagine it is less so than hoping that a high-drag device deploys on time."

"A parachute?" the Calmt said. "Not opening?" She nodded. "It happens very rarely these days."

"Very rarely for us, too," Cin replied. "And, of course, mostly we use remotes for the momentum exchange."

"And you will train our people to perform this momentum exchange?"

"So I understand," Cin said. When she saw the questioning look in the other's eyes, she added, "It makes more sense to have a trained planetary crew. With momentum exchange, half will end up with only orbital energy while half will have ship energy."

"I'm sorry, I don't understand."

"We use the momentum exchange between our incoming cargo and the outgoing cargo to slow down the incoming and speed up the outgoing—it saves energy and costs."

"And the bouncers bounce off the planet to change direction," the Calmt added.

"With equal masses, we can slow down and speed up equal cargoes with a ninety-five percent efficiency."

"Why not just have the cargo containers bounce themselves?"

Cin shook her head. "There is too little control. It works best when we can provide small changes in momentum."

"And it looks cool with all those flames in the atmosphere," the Calmt said with a grin.

"And it looks cool," Cin agreed, choosing not to argue the point.

"All orbital transfers involve momentum exchange in one way or another," the Calmt said. She nodded toward Cin. "Your way is elegant and efficient, if a little odd to those of us with lesser technologies."

"A lot of time was spent developing the momentum exchange," Cin said.

"The advantage of the smaller exchanges is a lower acceleration, less stressful to certain cargo," the Calmt observed.

"Yes."

"The human body can withstand accelerations many times that of normal gravity," the Calmt continued. Arwon was a nearly Standard planet, with a gravity just a bit below that of ancient Earth.

"True," Cin agreed, wondering where the woman was going with the conversation.

"Is the Calmt learning much?" a male voice spoke from Cin's far side. Cin turned to face a taller, handsome man dressed in clothes that were considered fine on the planet below. To Cin's eye, they were boorish and overbearing.

"Yes, I am, Your Holiness," the Calmt said, speaking around Cin.

"Good," His Holiness replied. To Cin he said, "The Calmt is expected to learn much."

"There are many things in which I am ignorant," the Calmt agreed. To Cin she explained, "My people only recently were integrated into the Greater Whole."

Cin gave her a blank look while sending a silent query to *Valrise*.

The most recent acquisition by the large, unifying government, Valrise explained. *There was fighting and many casualties. The details are suppressed.*

Can you not discover more? Cin asked through her link.

I am trying, Valrise replied. *But I must consider the impact on trade if my inquiries are discovered.*

"In fact, Your Holiness, I was hoping that I might participate in the training for our bouncers," the Calmt said, glancing hopefully to Cin.

"That is not—" from Cin clashed with:

"I don't see—" from His Holiness.

The two exchanged looks of chagrin. Cin motioned for His Holiness to speak with a polite, "I'm sorry."

His Holiness accepted it as no more than his due and continued, "I don't see why you would prosper from such training."

"I am told it is very risky," the Calmt said, "and I should like to know the risks myself so that I can best describe them to others."

Cin bit back her immediate response: It wasn't risky, only delicate. Instead, she said, "Most of us have implants and years of training, Your Holiness."

The Calmt gave Cin a half-pleading, half-grateful look. Cin

got the impression that the Calmt *wanted* His Holiness to be concerned by the risks.

She is offering to risk death for the Greater Whole? Cin asked *Valrise* through their link.

So it would appear, Valrise agreed. Cin was troubled by this notion; *Valrise* seemed just as troubled.

"If you wish it, I will consult with the Council," His Holiness said to the Calmt. To Cin, he said, "Can she learn without your 'implants'?"

"All the recruits will start without implants," *Valrise* spoke up. "I'm sure that, if it is acceptable to you, we would have no difficulty training the Calmt."

His Holiness jerked around, trying to find the owner of the voice.

"That was *Valrise*, our ship," Cin explained.

His Holiness made a warding gesture before quickly composing himself. "I had forgotten that your ship talks. It must be quite alarming."

"We get used to it," Cin said. "She has saved more than one life with her quick warnings."

"Really?" His Holiness said. His brows furrowed. "And why would she?"

"Trained personnel are hard to replace," Cin said.

His Holiness considered that. "I suppose that is so."

He turned his attention to the Calmt. "If you are finished with your interrogation of the technician, then we should return to our meeting."

The Calmt gave Cin an apologetic look before nodding to His Holiness. "If you will lead me, Your Holiness."

"Always," His Holiness said with a grim smile.

The skating event continued. Cin found her appointed seat—some distance from the Arwonese delegation but still close enough that she could see them. There were six members—and the Calmt was the only female among them. She was also the only one of her coloration and ethnic look. The others talked more amongst themselves than they did with the Calmt. She seemed very much the outsider. Cin thought she was both lonely and scared.

You have an observation? Valrise spoke on their link.

Cin relayed what she'd been thinking.

I cannot determine whether their behavior is ostracism because of her race or her gender, Valrise said. *Trade is the best way forward.*

That was what *Valrise* always said. It was what everyone on the ship believed—there were centuries of data to confirm the belief.

Trade always started slowly with a recontact world. First came raw materials and handcrafted artifacts, in return for which the ship would trade increased technological abilities. Slowly the world would be brought up to the standards of the rest of the galaxy. *Valrise* would generate goodwill and continue to profit from all trade. It made the ship and her crew independent of any one world for their needs. "Traders to the stars," that was their credo: Cin believed in it firmly.

This is your first recontact, Valrise observed. *It is not unusual to have misgivings and concerns.*

Do you also have concerns? Cin asked.

Always, the ship replied in laughing tones. *Will it ease you to train the Calmt yourself?*

Cin thought it over, glanced toward the small woman in the distance and nodded, knowing that *Valrise* would correctly interpret her nonverbal agreement.

The Arwonese were served lunch. Cin had already eaten and did not get a meal. She saw the Calmt watching her. Cin smiled her way. The Calmt gestured her over. Cin rose and came to her, nodding an acknowledgement to Captain Merriwether who had been briefing the Customers on all the events in the ice rink.

"Why are you not eating?" His Holiness said when he saw Cin approach.

"I've already eaten, Your Holiness," Cin told him.

"We consider it important that we break bread together," His Holiness said.

"If that will please you," Cin said. "My tastes may be different from yours. Would you like me to eat from your plate or get another?"

"I have a taster for that," His Holiness said. He glanced to the Calmt. "You may eat from her plate."

Cin started to object but the Calmt merely broke a piece of her bread and passed it, wordlessly, to her. Cin smiled at her, took it and broke it in half, proffering one half back to the Calmt while popping the other half into her mouth.

The Calmt smiled as she took the other half and ate it.

"There is a hint of cinnamon in the bread," Cin said when she'd finished her piece. "On *Valrise* we often consume cinnamon raisin toast."

"We are not familiar with 'raisins,'" the Calmt said. "What are they?"

"They are dried grapes," Cin said. Quickly, she added, "Grapes are grown on vines in moderate climates. Wine is also made from them."

"We have moderate climes but no grapes," the Calmt told her sadly. She gave her a wistful look. "Would it be possible to try some of this bread?"

"Of course," Cin said. She sent in a silent request to *Valrise*. The ship's response made Cin blush. "Oh, I'm so sorry! *Valrise* informs me that we don't know how your metabolism would handle raisins!"

"Are you refusing food to the Calmt?" His Holiness asked in a tone that was more amused than affronted.

"I am merely concerned that your metabolism may not be able to handle it," Cin said. "Your people might have an adverse reaction."

"If anyone should be the judge, it is the Calmt," His Holiness allowed. He turned to the smaller woman. "What say you?"

"I am willing to try," the Calmt said. She gave Cin a reassuring look. "I have recently discovered that I am able to stomach many things, some more unpleasant than others." His Holiness jerked at her last words and glowered at her.

Cin decided that the best way to defuse the tension was to accept the Calmt's assurances. A small tray of sliced bread, some toasted, some not, quickly arrived on Cin's table.

"What servant brought that?" His Holiness asked, glancing around in surprise.

"Our technology allows us to move things easily," Cin lied. She knew it was too early to mention their nanotechnology—it would appear to be too much like magic. And Cin knew that these people had only recently stopped persecuting "witches."

The Calmt reached for a slice.

"This is bread, this is toasted bread," Cin said, pointing to the two sides of the tray. She picked up a piece of bread, broke it in half and passed one portion to the Calmt. The Calmt eyed it carefully, sniffed it, and pulled out one of the raisins.

"It smells sweet."

"It is," Cin agreed, taking a bite of her piece. "The raisins are dried grapes. As they dry the sugar content increases."

The Calmt pulled off one raisin. She eyed it for a moment and then placed it into her mouth. "It is soft and squishy," the Calmt said. Cin was surprised at her choice of the word "squishy"—it seemed out of character with her serious expression.

Before swallowing, the Calmt took a small bite of the bread. She smiled at Cin when she swallowed her piece. "This is excellent!"

"I should like to try," His Holiness said, extending a hand toward the tray.

Cin passed it over and shortly, all the bread and toast had been consumed by the emissaries. Another, larger, tray appeared.

"Is this available for trade?" one of the men asked. Cin took in his sharp features and decided that this person was a trader or merchant like herself.

"I am sure something could be arranged," Cin said, using the time-old circumlocution.

Well done, Cin, *we'll make a trader out of you, yet! Valrise* teased through the implant.

I prefer bouncing, Cin replied with mock-seriousness. The ship's response was a snort of laughter.

Out loud, *Valrise* said, "We have a limited supply of raisins on board this trip. I'm sure we could bring more in the future."

"And how long would that be?" the trader asked. "Also, how many of these 'raisins' do you have for trade?"

"Everything is for trade...at the right price," *Valrise* replied.

His Holiness shot the trader a quelling look. "Our trade is precious, Van Domit, as I'm sure you know."

Van Domit gave His Holiness a deep nod. "I meant no disrespect. But I believe a trade in this commodity would be to the profit of all, Your Holiness."

His Holiness considered the matter and then said, "And what would be required for this trade?"

He looked toward Captain Merriwether. Cin bit back a smile: The captain was old and not altogether "there" anymore; everyone knew that *Valrise* was in charge.

"I would ask the ship," the captain replied, jerking a finger up toward the ceiling.

"Indeed," His Holiness replied, seeming out of sorts. He took a deep breath, sighed and glanced upwards. "Could we arrange a trade, ship?"

"I believe we could," *Valrise* replied. "In fact, if you desire, we could go one better."

"Better?"

"I understand that you have a large area on your southern continent that has recently become uninhabited," *Valrise* said. Cin saw the Calmt jerk in reaction to her words. "We would be willing to lease part of that land and use it for agricultural purposes. Our sensors indicate that it would be perfect for growing grapes."

"The grapes that become raisins?" the Calmt asked with alacrity.

"The very same."

"And you would trade them to us?" Van Domit asked.

"Of course," *Valrise* replied. "We would train your people in their planting and harvesting, naturally." There was a slight pause. "Would that be acceptable?"

Van Domit looked ready to shout with glee but controlled himself with some effort to give His Holiness a pleading glance. "It could profit the Church."

"We have some lands there that might serve," His Holiness allowed. He waved the matter aside. "It is a matter for later, perhaps."

"Trade is not best made waiting," Cin said, recalling an old saw among the ship's crew.

His Holiness gave her a sharp look.

"What I mean, Your Holiness, is that all too often good ideas are lost in the crush of events," Cin said quickly. "It seems that we have discovered another source of trade and have come to an agreement. Why not settle it now?"

"It would profit to do so," Van Domit said in agreement. Hastily he added, "Your Holiness."

His Holiness pursed his lips. Finally, he nodded. "Very well," he said. He glanced to Van Domit. "I shall leave the matter in your hands. I trust the Church will receive all due accommodation."

"But of course, Your Holiness," Van Domit said obsequiously, clasping his hands in front of him and rendering His Holiness a respectful half-bow. "Your will shall govern our Council."

Cin couldn't tell if the man was being sarcastic. She knew from *Valrise*'s briefing that His Holiness had the final say in the Arwonese Council. His Holiness seemed not to care, deliberately taking another slice of toast and carefully consuming it.

When he was finished, he turned his attention back to the display on ice.

Cin had to admit that the skaters were good—very good. The downsiders had had only a few days to train on the two gravity fields but they were the best of their world—and it showed. The ship's crew had welcomed them openly and they had trained together extensively. Now, they worked together in a display of cooperation that spoke well for future efforts, effortlessly teaming up to provide a whip line of five skaters and whirling themselves up and over to the inverted gravity field and then back down again. At one point they flipped the line so that one of them floated in the null gravity while the pair on either side flipped over and over as the central anchor precessed in a circle with them.

"This is amazing!" His Holiness declared. He turned to Cin. "Is this not the greatest wonder your ship can offer?"

"It is one of them, Your Holiness," Cin replied.

"One of them?" Van Domit repeated in surprise.

"Is there any better?" His Holiness asked.

"Those of us who ride the atmospheres in our bouncers think *that* is the best," Cin replied with a dimpled smile.

"I should understand how you, being one of these 'bouncers,' might feel that way," His Holiness returned. "But is it not true that it is a very inefficient way of changing orbits?"

"Inefficient?"

"Wouldn't your antigravity provide you with a better method?" Van Domit added.

"We could indeed use antigravity," Cin conceded. "But over the centuries we have found that using the bouncers and momentum exchange is more efficient—"

"How so?" Van Domit interjected.

"Antigravity is not perfectly efficient, which means that more energy is required to create it," Cin said. "We are a trading ship and anything that cuts into our profits is to be avoided. Over the centuries we discovered that exchanging momentum from the inbound cargo to the outbound cargo was most efficient."

"Difficult, too, no doubt," Van Domit said.

Cin nodded. "It takes a lot of calculations and a little patience— we could move cargo faster with antigravity but, seeing as our ship is inbound to your world and will pass it on its outbound course to our next destination, we have enough time to prepare."

"You say 'patience' but you also talk of a very tight schedule of momentum exchanges," the Calmt said now.

"I do," Cin agreed with a grin. "We've plenty of time now to practice and prepare but we'll have to be spot-on for the exchanges to work."

"And if they don't?" His Holiness asked.

"We always build an amount of redundancy into our schedule," Cin said. "While it's very rare for us to 'botch' a set of maneuvers, we *can* recover."

"If profit is so desirable, how do you explain shuttling us all the way out to the edge of our solar system to meet with you on your ship?" His Holiness asked.

"That's customer service," Captain Merriwether said. He waved a hand toward the skaters. "We don't want merely to haul one cargo for you; we hope to become trader partners for a long time."

"And to teach us your bounce technology?" Van Domit said.

"Of course," Captain Merriwether agreed. "We consider it one of our greater gifts."

"*This*, however, shows a reverence for the Lord," His Holiness said, jabbing his hand toward the skaters, "through his gifts of our bodies and our minds."

"Indeed," Cin agreed. "For myself, I often find much to marvel upon as I bounce through the red of another world's air."

"The Calmt, then, shall try this and give us her opinion on the matter," His Holiness said with a curt nod toward the Calmt Prime. The Calmt nodded back, then gave Cin a grateful look. Cin returned it with a smile of her own.

Out of the twelve Arwonese assigned to bounce training, only three were women—and all bore the same features as the Calmt Prime. Cin learned that they were Calmterians, named after their continent on Arwon. The Calmt Prime was their acknowledged leader, although Cin caught some strange undertones in the way the other two women, Mira and Sikar, spoke of her. It was like their respect for their prime was conditional. Whatever it was, it did not affect their training.

It bothered Cin, however, that the three women chose to keep to themselves and were reluctant to train with the men. She and *Valrise* attributed at least part of that to their gender—it was clear that the males considered them of lesser value even when they demonstrated their technical prowess for all to see.

Valrise's trajectory would pass Arwon in a week, but in the meantime, they were too far from Arwon for the bouncers to train in the planet's atmosphere. *Valrise* dispatched *Lewrys*, the same transport that had been used to bring the Arwonese out to the ship, to bring the bouncers in for practice.

Cin had let Mira take the controls of *Lewrys*, knowing that *Valrise* herself would keep a careful eye on the trainee. Coklon—one of the more sensible men, in Cin's opinion—sat in the copilot's chair. The two worked together well, if without any real chemistry.

Chemistry! Cin chided herself. But that was the best way to reflect upon their interactions. Mira and Sikar worked together flawlessly, slightly less well when paired with the Calmt—there was a deference, almost a reluctance in their actions—but excellently when tripled together.

"Today we will practice bouncing," Cin said. "You've all done well in the simulators but now you're going to find that the real thing is an entirely different matter." She paused. "Some of you will discover that bouncing is not for you. There is no shame in that. If you find yourself overwhelmed, we will recover you automatically, you need not fear."

The men all assured her that they would not "chicken out," as they said on Arwon.

"First up, Coklon and Alvar," Cin said.

Cin followed Coklon and Alvar on her monitors as they performed their preflight checks, nodding as they proceeded normally. She threw in a slight curve for both of them and they caught it—in the case of Coklon it was a failing transceiver, for Alvar it was a low battery. Both "failures" were acknowledged and corrected.

"Prepare to undock in three . . . two . . . one . . . go!" Cin called. They were under manual control—more testing. The two men undocked within milliseconds of one another—almost as good as regular crew. The bouncer rotated and set up a course toward the atmosphere.

Cin double-checked their vectors. Alvar was slightly off. She frowned, wondering whether she would tell him, but before she'd made her decision, his voice chirped, "Vector correction, azimuth increment zero point five."

She checked the trajectory and said, "Vector correction acknowledged and confirmed." A moment later, she added, "Good catch."

A burst of static that might have included a snort of laughter came back to her. Alvar was cool, levelheaded, and known for a fairly dry sense of humor.

Coklon, meanwhile, was set up for a textbook insertion. He'd hit his mark and then bounce back up and away toward their rendezvous point with the test cargo vessel. They would impact it, exchange their momentum and return for another bounce back to *Lewrys.*

Cin double-checked with *Valrise,* who agreed. Cin told the two bouncers, "You are 'go' for deorbit and bounce."

"Coklon copies 'go' for deorbit and bounce."

"Alvar copies 'go' for deorbit and bounce."

The two bouncers slowed and started their descent to the planet's atmosphere. Their velocity and trajectory were too shallow for reentry; instead they would skim off the atmosphere and rebound back into space on a new trajectory with reduced energy. When they hit their cargo target, they would exchange energy and fall back into the atmosphere for another bounce.

"Atmosphere," Coklon reported right on schedule. Cin's brows creased as she looked at Alvar's displays. The man said nothing. She checked his medical readouts: heart rate high, pulse high, breath rate high—he was cracking up.

"Alvar, report," Cin called on her private frequency. She waited seconds beyond his response time. "Alvar, I know what you're feeling. We're on a private connection; no one can hear you but me. Just take a breath. Take a breath and press your transmit button."

"I'm going to burn up!" Alvar shouted a moment later. "It's too hot!"

"Alvar, try a sip of water," Cin replied. She knew that the bouncer's skin temperature was normal, its internal temperature nominal—the "heat" was Alvar's fear. She waited, hoping he would follow her orders. "Alvar, drink. The water's cool."

A moment later, she saw that he had sipped.

"It's too hot, I'm burning up!"

"Do you want recovery?" Cin said. "Do you want me to bring you back?"

"It's all red outside! Glowing! There are flames! I can see

flames!" Cin checked her readouts and brought up visuals from inside the bouncer.

"There are no flames, Alvar," Cin told him calmly. "What you're seeing is normal for your altitude and speed. You are in the green, textbook."

"I'm dying, I'm dying!" Alvar wailed. Readouts showed his heart rate climbing alarmingly.

"I'm going to pull you out, Alvar," Cin said.

"No! No, if you do that, I'll die!" the man's voice roared in her headphones.

"You're completely safe," Cin said. She readied the abort sequence but held off for a moment.

"No, I'm burning up! And if I go back, I'll die! I know it!" He roared. His voice lowered. Pleading. "They'll kill me, don't you see?"

Cin hit the abort. On the all-call channel she reported, "Systems malfunction on Alvar's bouncer, recovery initiated."

"NO, NO, NO!" Alvar cried on their private frequency.

"I said it was a systems malfunction," Cin assured him. "Not your fault."

"No! Let me go!" Alvar cried. "I can handle it!"

An alarm wailed suddenly in the shuttle's cabin. *Manual override! Valrise* warned even as the shuttle's com announced, "Manual override on bouncer Alvar, manual override."

"Alvar, reengage the controls," Cin said. She checked her readouts. "You're going off course."

No response.

"Alvar! Alvar, you must let the computer take control—you're off course!" Cin said even as *Lewrys'* telemetry indicated that Alvar had engaged his thrusters...in the wrong direction.

"Warning, systems anomaly, temperature variance increasing," the autonomous telemetry for Alvar's bouncer reported.

"Initiate emergency recovery," Cin ordered. She knew better—the first thing she had trained the Arwonese to do was to disengage the automatic systems in case of emergency. While Alvar had been operating under manual control, he had to purposely cut the circuit to disengage the automatics.

"Systems off-line," the bouncer's system reported. Alvar had clearly retained that much knowledge in his panic. "Hull breach imminent."

"Alvar, you've got to change your course!" Cin shouted. "You're burning up!"

"It's beautiful!" Alvar's voice came back. "It's so beautiful!"

"Contact lost," *Lewrys'* tracking systems announced. "Debris detected."

Alvar's bouncer disintegrated, Valrise reported. Cin immediately checked on Coklon: He was still on profile.

"Coklon, report," Cin said.

"Everything is nominal," the Arwonese man replied laconically. "It's beautiful down here!"

"What happened to Alvar?" Mira shouted loud enough to be heard through Cin's headphones.

Valrise? Cin asked through her link.

The bouncer disintegrated, Valrise replied. *We shall record it as a malfunction.*

Cin pulled off her headphones and said to Mira, "Something went wrong with his bouncer."

"Wrong? What happened?"

"He didn't make it," Cin told her.

Captain Merriwether handled the discussion. "We are too close to change our delivery method," he told His Holiness. "We can abort the transfer."

"And if you do?"

"We will return when we can," Captain Merriwether said. "Probably in three years or so."

"This is your fault, Captain," His Holiness replied sternly. "You have admitted that there are other ways to deliver cargo and affect trade."

"Not as efficient, therefore more costly," Captain Merriwether said. "We've found that this is the best for long-term trade."

"Perhaps we don't want your long-term trade," His Holiness replied.

"For a fee, we are willing to announce your presence to other traders," Captain Merriwether returned calmly.

"For a fee," His Holiness sneered. Someone off camera spoke in urgent tones but Captain Merriwether did not hear what was said. The expression of His Holiness altered. "How often are your bouncers destroyed like this? Surely you lose a lot of your own crew in such accidents?"

"We experience a mechanical difficulty about once every one hundred thousand flight hours," Captain Merriwether said. He nodded toward the camera. "You have that information in your initial contact packet and signed the contract acknowledging the possibility."

"One of our *men* died, Captain," His Holiness snapped back. "We now wonder whether your contract has any merit."

"That is your decision, Your Holiness," Captain Merriwether replied calmly. "If you wish us to deliver cargo, we must continue the training immediately."

"And what if we leave the operation solely to your crew?"

"Then we will abort the transfer," Captain Merriwether said. "If it is of any aid, you may contact your remaining personnel. They have assured me that they understand the risks and are willing to continue their training."

"All of them?" His Holiness asked.

"All of them."

His Holiness mulled on that. After a moment, he nodded. "I should like to speak with them in private. If they agree, I will approve."

"Very well," Captain Merriwether said.

"Now, as we all know, we suffered a loss but we have agreed to continue this mission," the Calmt Prime said as she looked over the Arwonese bouncers. "We have been training in simulators since the accident; today we are back in real ships." She nodded toward Cin. "Cin will accompany us on our bounces while the *Valrise* will provide additional support." She paused and glanced at Mira and Sikar, who gave her reassuring nods. Then she looked to the men.

"Let's get to it, time's wasting!" Coklon declared.

"I shall go first," the Calmt declared. The others gave her shocked looks. "It is my right and my duty."

Cin said nothing, gesturing for the Calmt to enter her bouncer. Cin triple-checked the bouncer's systems: As expected, all were perfect. She checked her own unit—built for her and to her specifications. Her bouncer was named *Terra*, after mankind's homeworld. Cin had promised herself that, one day, she would bounce on the homeworld itself and the craft's name reflected that promise.

The two undocked from *Lewrys* and commenced their deorbit burn.

The Calmt is requesting a private channel, Valrise informed Cin.

Fine, Cin replied, toggling a connection. "You wish to speak in private?"

"I do," the Calmt said. "I have been thinking—"

"We could abort, if you wish," Cin said.

"Not that," the Calmt said. "I have been thinking about Alvar."

"We have determined the cause of the error, and I can assure you that your bouncer is operating nominally," Cin told her.

"Of course," the Calmt replied in a smug tone. "I am convinced that the error has been corrected. Alvar is not at the controls."

"What do you mean?"

"He caused the accident, didn't he?" the Calmt said. When Cin didn't immediately reply, she continued, "He panicked and cut out the fail-safes."

"What makes you say this?"

"There is something you don't know about our world," the Calmt said.

"There are many things I don't know about your world," Cin replied.

"Alvar panicked, didn't he?"

"*Valrise* reported a systems malfunction," Cin temporized.

"We suffered a major famine and our people are starving," the Calmt said. "Those who were selected for the bouncers were expected to succeed. We need your trade if we are to survive. His Holiness made it quite clear that anyone who did not learn to use your bouncers was 'a mouth wasting food'—his own words. So when Alvar panicked, he knew that he would die. One way or another."

"Why did you not say there was a famine?" Cin said. "*Valrise* is committed to humanitarian aid."

"It is a very *select* famine," the Calmt replied. "Those His Holiness considers unworthy are those that starve."

"Are your people among them?"

"My people are *most* of them," the Calmt said. "We are the ones most recently 'assimilated into the Greater Whole.' Are you familiar with the concept of a 'scorched earth policy'?"

I am, Valrise said to Cin, relaying relevant images. Cin fought an urge to vomit.

"Why did you not contact us directly?"

"We could not; we had no transmitters," the Calmt replied. "I was elected among my people to join the Greater Whole, to renounce my kind in the hope that I might protect them from within." She added bitterly. "In that, I failed."

"So why do you continue to work with His Holiness?"

"Who says I am?" the Calmt replied.

Cin checked her readouts and waited. The Calmt said, "Coming up on bounce in five minutes."

"Watch it, it'll be beautiful," Cin said almost automatically. She'd been watching the outside of her bouncer lighten and had noted the first pink rays of atmospheric heating color the nose of her ship. Unlike the Calmt, Cin wore nothing when she dove. Her body was encased in an air-replacement fluid; it filled her lungs, allowing her to perform higher-gee maneuvers than the normal bouncers could. Like all the bouncers, her ship's hull was transparent, unlike those for the Arwonese. Her instrument panel was also transparent, marked only by the small specks of the highly integrated circuits which comprised her computer controls. Cin's craft had a slightly lower thermal insulation, allowing her to feel the temperature rise as her bouncer sliced through the atmosphere. She—and all the other bouncers in the crew—liked the notion of being able to "feel" the air around them.

They bounced. Minutes later they impacted the target cargo vessel and reversed course.

"Amazing!" the Calmt declared as they returned to the red depths of the atmosphere. "Absolutely amazing!"

Cin met the Calmt as she emerged from her bouncer. Cin had already donned a shipsuit to avoid offending the Arwonese woman.

"Your skin is glowing," the Calmt declared, raising a hand to her own face. "Is mine like that?"

"No," Cin said. "When I bounce, I wear no clothing."

"Isn't that dangerous?"

"My bouncer is all I need to wear," Cin said. "I like to feel the heat of the air on my skin."

The Calmt shook her head in wonder. "I don't think that is too strange for my tastes." She waved the topic aside. "How are the others doing?"

"They are within tolerance," Cin replied, having checked in with *Valrise.*

"I'm sure His Holiness will be pleased," the Calmt observed tartly. "Will we be able to proceed, then, with the transfer?"

"We have to make up for lost training time but I believe, with no further difficulties, it will be possible."

"And the remotes will help," the Calmt said.

"Training with the remotes will become the first priority," Cin said. "The remote bouncers are always the primary form of momentum exchange."

"The human bouncers are just for fun," the Calmt guessed.

"Not just for fun," Cin corrected. "However it is true that we work best in managing the remotes."

"Why couldn't the ship do it herself?"

"It is a question of proximity," Cin said. "*Valrise* will not be close enough to provide immediate responses."

"But isn't this all rather predetermined?" the Calmt asked. "I mean, isn't it just a matter of physics?"

"Almost," Cin replied. "Atmospheric turbulence means that not even the most advanced calculations will be completely predictive. There is an element of chance that must be managed." Cin added, "And sometimes the masses of the cargo containers are not as accurate as desired. We know how to handle that."

"And will we learn to manage that?"

"Mostly that will be up to the crew bouncers," Cin said. She caught the Calmt's look and added, "As you get more experienced, we will bring you more closely into the operation."

"You mean, the next time you come back."

Cin still had misgivings about the proficiency of the Arwonese bouncers.

"There is something going on," Cin informed *Valrise*. "I see too many strange looks passing between them, particularly the women."

Their performance is above average, Valrise replied. *And we must make the trade, particularly after losing one of their own.*

Cin frowned but nodded in troubled agreement. The integrity of the ship was at stake.

Keep an eye on things; let me know if anything untoward occurs, Valrise added.

"And you do the same," Cin replied.

They are planning several low-level launches, setting up the

GPS system we recommended, Valrise reported. *Beyond that, their space efforts are limited strictly to the cargo exchange.*

"We'll know one way or another the next five hours," Cin said aloud, glancing at the automated countdown timer.

Lewrys and *Marys*—both named after revered and long-dead crew members—were prepped for launch.

Cin was aboard *Lewrys*, ferrying the downside crew—mostly the Arwonese—to their destination. Goroba had the rest of the bouncers with him on *Marys*. Cin had no concerns about the crew bouncers, so she'd suggested that she ride with the Arwonese. *Valrise* had agreed.

With Emery as pilot, the shuttle was cramped with the eleven trainees and the two crew. Its racks carried a full complement of crewed and remote bouncers—forty in all.

Fortunately, they didn't have far to go. Arwon was now a large ball filling their port side almost completely.

"Once again, tell me the drill," Cin said to the collected Arwonese bouncers.

"We drop anti-orbit, hit our cargo, bounce back, and hit the inbound cargo," Coklon said with a grin. "Repeat and continue until sufficient momentum is exchanged, then rendezvous back here."

"Where *Valrise* will place an orbiting station which will include a reentry vehicle," the Calmt finished, glancing at her two fellow women. Cin's brows creased as she caught their looks: It was that same disturbing look she'd seen before.

"*Arwon I,*" Coklon said in agreement. "The Blessing has already been bestowed by His Holiness."

Cin kept her face blank. She'd had little time to learn more about the politics of Arwon but what she had learned—through her training and in conversations with other crew—was enough to make her wary of religious oligarchies.

Coklon snapped his fingers in an expression of the ease of the task facing them.

"Prepare your bouncers," Cin said. To Emery she said, "Deploy the remotes at your discretion."

"I'll deploy 'em when you launch," Emery replied with a chuckle. He and Cin had once been lovers; they still worked with the easy comradery of two people who had shared thinking.

"Just don't be late!" Cin teased. She moved down the corridor to the hatch and her bouncer. Out of sight of the Arwonese, she gladly shucked her shipsuit before wriggling through the hatchway and onto her craft.

"Sealing," Cin reported, activating the controls. Her bouncer, *Terra*, responded eagerly, like some ancient steed awaiting her commands. Cin closed her eyes, engaged the atmospheric controls and waited as the cool, slightly damp gas rose from the reservoirs on board the *Lewrys* to cover her completely. She took the breath that she always hated as the gas rose to the level of her lips. It rose over her head quickly and Cin forced herself to inhale it. For a moment she fought panic as her lungs instinctively tried to expel it. The gas condensed slowly around and inside her, turning to a body-temperature liquid that encased her protectively within *Terra*'s super-tech transparent hull.

"Life support," Cin said.

Life support one hundred five percent, Terra responded through Cin's link. The bouncer was considered operational with life support as low as ninety percent, but *Valrise* insisted that there be a margin for error in all normal operations. With bouncers, normal was a rare condition, *Valrise* had once explained to Cin, alluding to the general crew belief that anyone who would expose themselves to a hard atmosphere in extreme conditions was very much outside the norm. Crazy.

Cin completed the rest of her systems checks and turned her attention to the earthers preparing beside her.

She was not surprised to learn that all the women had chosen to leave their shipsuits behind. What did surprise her was that Coklon and Batric had also "gone native." For herself, Cin couldn't imagine bouncing in something as constricting as a suit. The gel-liquid that surrounded her operated with less friction than a suit, making her movements just that much quicker and efficient. Early in her training—and anytime she was training worlders—she had to wear a shipsuit through a bounce; no matter how transpired, she never liked it.

"All systems nominal, we are coming up on insertion," Emery reported just as Cin had come to the same conclusion.

"Call, Calmt," Cin said. She was letting the Calmt call the orders.

"*Valrise, Valrise*, this is LEO bounce team, request insertion," the Calmt said with the clipped precision of a well-trained professional.

"LEO bounce team, I copy and confirm that you are ready for insertion," *Valrise* replied over the regular comms.

"They're good," Cin said, emphasizing only that she thought the team was ready.

"LEO team, you are cleared for insertion," *Valrise* said on the comms. "Wait one, I have a transmission from downside."

"Waiting," the Calmt replied, her tone notched up in what sounded like worry.

"Stand by for His Holiness," a voice, modulated by transmission from the distance, through an atmosphere, and reflecting somewhat inferior electronics, spoke up.

"We are ready, Your Holiness," the Calmt said.

"Godspeed," His Holiness' voice came back. "May this be the beginning of a great new adventure."

"Thank you," *Valrise* replied. "We look forward to continued trade."

"Do we have your permission to commence the operation, Holiness?" Coklon asked.

What? Cin shot to *Valrise*.

"You are authorized to proceed," His Holiness responded. "Godspeed in all your efforts."

"Coming up on insertion," Emery said with a note of worry in his voice. Cin checked her readouts: If they waited too much longer, they'd have to abort to the next orbit.

Cin? Valrise asked.

"What if we don't go?"

Trade would be impacted, Valrise said.

"Then go," Cin said.

"LEO team, separate," *Valrise* spoke over the comm. To Cin, she added, *Keep your eyes open.*

Cin watched as the team separated in order. Cin went last, the Calmt just before her.

"Insertion," the Calmt ordered when everyone confirmed their positions.

"Initiating," Cin responded when her mark showed.

Little pinpricks lit the sky around her as the bouncers began their descent into the planet's gravity well.

"Cargo marks," the Calmt called out.

"Roger, targeting cargo," voices came back in confirmation.

"Drone release," Emery called.

"Roger," the Calmt replied. Cin double-checked the telemetry. "Drones on course."

In addition to the eleven bouncers there were now thirty-three remotes en route—allowing the combined bouncers to manage the eleven upbound cargo containers which were themselves matched by eleven containers coming from *Valrise*. They were not matched for mass. The outbound containers were far more massive than those coming from *Valrise*. That reflected the greater value of the incoming cargo—gram for gram it was worth nearly a thousand times more.

The maneuvers—bounces—were fairly straightforward: Cin and the others would change the initial circular orbits of the containers into highly elliptical orbits, then turn the orbits around—this was possible because at an orbit's apogee the velocity of the object was nothing—minimizing the energy required to alter its trajectory.

So the LEO crew would bounce the outbound cargo up, then exchange momentum with the inbound cargo—slowing it down by speeding themselves up—and repeat the process.

The process became involved because of the relative masses and the need to make the momentum-exchanging impacts at sufficiently low gravities.

Bouncers could safely handle ten gravities—Cin herself had tested to twenty. The containers were often much more fragile: taking only three gees maximum. This was particularly true for the planet-built outbound containers. Included in the value of the exchange was the higher-tech of *Valrise*'s containers—containers that were standardized in the Trade.

According to preliminary calculations, it would take twenty-three exchanges to complete the momentum translation.

The first impacts would be glancing blows as the bouncers dropped to the atmosphere.

Cin double-checked the trajectories—the earthers had to hit center of mass in two dimensions to avoid imparting any spin or yaw to the containers. The remotes could help in correcting any errors. And, in this first instance, the rules of physics were the only rules to be considered. Once bouncers hit the turbulent atmosphere of the planet, inconsistencies would be introduced.

"Impact in three...two...one..." Coklon called out. His bouncer and the three remotes were the first scheduled to hit a container. "Impact!"

On profile, Valrise confirmed. Coklon's ship rebounded from the container which seemed unaffected by the multiple impacts. Cin's telemetry showed otherwise and she added her voice to the others congratulating the Arwonese man on his first bounce.

One by one the others hit, all on profile. Cin grunted as *Terra* impacted on the last, and most difficultly placed, container.

Off profile! Valrise reported even as Cin's telemetry flashed red. *There is a mass discrepancy! Center of gravity also does not correspond.*

"We got a bad bounce," Emery reported from *Lewrys*. "Recomputing." On a private channel he said to Cin, "What did you do, woman?"

"Not me," Cin replied. "Something's off."

"Off by a tonne," Emery reported. "The ship's on it; she's talking with the downsiders."

"Recomputing bounce," Cin responded, toggling her computer interface. Fortunately, the bounce was not too far off profile; she could easily correct. She was glad it was her and not one of the Arwonese: They would have been dismayed by the prospect of a deep bounce.

"Got your will in order?" Emery teased over their private link.

"Huh!" Cin snorted derisively. It was an old joke common between bouncers and pilots: a part of the rivalry between two "crazy" professions.

"Cin—" the Calmt called on a private link.

"We have communications from earthside—" Captain Merriwether reported at the same time.

Valrise?

Trouble, the ship responded.

"What is it?" Cin said to the Arwonese woman.

"Missile launch! Multiple missiles inbound!" Emery roared over the link. "Red watch, red watch, red watch! We are under attack! I repeat—"

His voice cut out at the same moment that Cin lost telemetry with *Lewrys*. Instinctively, Cin kicked her thrusters, twisting her vector and velocity at the same moment.

Valrise!

Her comm filled with the voice of His Holiness. "The unholy must be cleansed."

"What—?"

"The containers, we must save them!" the Calmt cried on her link.

Missile lock, armed, impact in two hundred seconds, Terra warned.

Cin kicked her thrusters to max, setting *Terra* to dive into the atmosphere.

"Follow me!" Cin called over the private link. She sent the same instructions to her remotes but only two responded. A quick check showed her that the other two had been destroyed.

As had all the bouncers save Mira's and the Calmt's.

"What's so important about the containers?" Cin shouted over her comm link.

Communications loss in ninety seconds, Terra warned, referring to the standard atmospheric plasma disturbances that occurred during a bounce into an atmosphere.

"My people," the Calmt replied.

"What?"

"There are two thousand women and girls on your container," the Calmt told her. "They are the last of my people."

"What?" Cin checked telemetry: All the other containers had been destroyed.

"Please, you must help us," the Calmt replied. "We had filled four of the containers with our people. This is the last one."

"But—"

"Our people were destroyed; they were going to be eliminated," Mira said on the private circuit. Brokered into the link by the Calmt, Cin surmised.

"We hid them," the Calmt replied.

"I thought you were part of the Greater Whole?"

"I pretended to betray my people," the Calmt said. "I sat on the Council while they were destroyed, trying to find some way to save them."

And now this, Cin thought.

"Why didn't you tell us?"

"And be discovered?" the Calmt asked. "Until my people were in space, there was no surety." She paused. "And even then..."

Debris fields calculated, Valrise added. *Prepare for updates.*

"Loss of signal!" Cin warned, as her ship started to glow pink with the air rushing around her.

You are on your own, Cin, Valrise told her sadly. *The safety of the ship and crew are paramount.*

"I know," Cin said.

Do your best, daughter.

The link went dead: LOSS OF SIGNAL.

"I'm going to lose your signal," Cin said to the Calmt and her compatriot. "Just take your bounce and we'll talk when we get signal again."

"What are you going to do?"

"Bounce," Cin told her simply. "Bounce your cargo to my ship."

"*Terra*, recompute with maximum gravity impacts for quickest transfer of cargo," Cin ordered her craft. "Ignore safety margins."

Around her, the air grew brighter. Cin took a moment to bask in the glory that was a world trying to destroy her. The temperature rose as *Terra* hit her perigee and then she bounced off the atmosphere. Back to the stars.

"—almt calling anyone, please respond!" the Calmt's voice came to her, stressed with fear and worry.

"This is Cin, I'm recomputing now," she assured the Calmt. "Is Mira still with you?"

"Yes, I'm here," another voice replied, sounding less stressed and more awed.

"Bounce well?" Cin asked.

"It was beautiful!"

"And all the missiles burnt out trying to follow us," the Calmt remarked.

"That was the plan!" Cin said. "Do they have more?"

"Possibly," the Calmt said. "But it will take them some time to rearm and launch."

"Okay," Cin said. She was silent for a moment. "Two thousand?"

"Maybe more."

Terra completed the calculations. Cin glanced at them: She'd expected nothing more.

"Mira, would you like to *really* bounce?" Cin asked.

"Will it save my people?"

"It's their only chance," Cin replied. "We're going to have to go deep pink."

"Deep pink?" Mira repeated in confusion.

"She means we're going to dive deep into the atmosphere," the Calmt replied. "How safe is that?"

"Not very," Cin replied. "*Terra* is sending your bouncers the data now."

"Twenty gravities!" Mira swore when she got the download. "Can anyone survive that?"

"I have," Cin said. She didn't mention that it had taken her a week to recover. In this case, it didn't matter.

"If we die—"

"We'll time the highest gees for last," Cin replied. "We should be fine until then."

There was a moment's silence as the other two absorbed her words. They had the calculations: They knew the price.

"Very well," the Calmt said. "I am prepared."

"So am I," Mira said. "My sister is on that container."

"Then we'll give her the best ride we can," Cin said.

I can see no other way, Valrise said through their link. *We will receive your gift, Cin, have no fear.*

"Thank you."

I'm launching remotes to aid you, Valrise added. *At the very least they may be able to complete your mission.*

"Good," Cin replied. After a moment, she added with a laugh, "Does this qualify as combat?"

Most certainly!

They bounced four more times. The third time, a new launch of missiles picked off Mira's bouncer before she could get deep into the atmosphere.

Valrise had launched countermeasures by then so the next array of missiles from the planet were destroyed.

Not that it mattered: There was now too little mass to complete the momentum exchange.

"There has to be a way," the Calmt cried over their link. "There has to be!"

"Follow me," Cin said.

"What can we do?"

"Our ships are built better than we are," Cin told her. "They can handle fifty gees. And they can handle higher temperatures. The computers will do it all for us."

"But my people!" the Calmt cried.

"We're exchanging momentum," Cin reminded her. "They'll only get a sharp nudge."

"To do this we must go to the depths of the atmosphere?" the Calmt asked.

"To do this, we must melt," Cin told her grimly.

The Calmt was silent for a long moment. Finally, she said, "It will be a hell of a ride, won't it?"

Cin grinned. "No one will ever see its like."

"And live to tell the tale," the Calmt chuckled in bitter agreement.

"Ready to see the white at the end of the pink?"

"Yes."

Valrise recorded it all. She recorded their breaths, their heart rates, their skin temperatures, their pain. Their screams.

From the depths of the atmosphere two fiery, glowing spheroids rose back into the sky to hit the last cargo container with all their light and energy.

It was enough, as Cin had calculated.

"Prepare a grapple and secure that cargo," Captain Merriwether said over the comm.

"Aye aye, sir," the cargomaster replied with a nontraditional military bearing.

"You must return now!" His Holiness called over the link. "You are in grave danger; the people on that container are escaped criminals."

"*You* are in grave danger," Merriwether replied. "You have committed an act of war on a civilian ship." He paused for a moment, glancing at his telemetry. "You have killed four of our crew and thousands of your people."

"Only eleven are mine," His Holiness returned acidly. "The rest are vermin."

"We are taking your 'vermin' with us," Captain Merriwether said. He cut the comm. *Jump,* he ordered *Valrise.*

Alarms sounded as the ship prepared to jump into hyperspace.

A moment later, *Valrise* entered the nothingness.

A month had passed since the jump from Arwon. The refugees had been settled on a number of worlds, some had petitioned to join *Valrise* as crew. Some had been accepted.

Captain Merriwether and *Valrise* conferred on the final reports.

"I got good reads on her all the way down," *Valrise* said out loud.

"Cin?" Merriwether asked. His brows furrowed. "What do you propose?"

"I'm going to make another," *Valrise* declared. "We've room in the infirmary, and in the growth tanks; her genetics are on file."

"You're not going to give her all those memories?" Captain Merriwether asked, aghast.

"No, of course not," *Valrise* replied.

"You don't just want to name a shuttle after her?" Captain Merriwether said. "After all, she was just a bouncer. Hardly irreplaceable."

"All five of them," *Valrise* said. Captain Merriwether gave her a questioning look. "We have genetics on the Arwonese."

"They didn't consent," Merriwether protested. Typically, cloning required the progenitor's consent.

"My authority," *Valrise* said.

Captain Merriwether thought for a moment. Nodded. "It's your ship, after all." After a slight pause, he added, "And what was her real name, the Calmt Prime?"

"Sorka," *Valrise* replied. "She'll be Sorka Arwon."

Three of the girls were smaller and darker than the fourth. But they were inseparable: born of the same pod at the same time, along with the boy.

Captain Merriwether visited them when he could; *Valrise* was always with them.

"And what's your name?" Captain Merriwether asked the fourth girl.

"I'm Cin," the girl replied proudly. "Cin *Valrise* the Second!"

Pageants of Humanity

Brent Roeder

Brent is currently a neuroscience PhD student researching how to restore damaged memory function. A lifelong geek, he enjoys writing sci-fi and fantasy to relax from work. Very occasionally he even remembers to finish a story. We hope you enjoy his vintage take on the future of what it means to be human.

"To all of our new viewers, welcome to our coverage of the two hundred and thirteenth decennial Exposition of Humanity!" the male host announced cheerily from where he sat next to a female companion. "I'm Boberto Lopez and this is my co-host, Cindellou Whoon."

"And to all of our returning viewers, welcome to our new channel," the female host added with a smile.

"For over two thousand years the Exposition of Humanity, or The Expo, as we like to say, has been helping humanity recognize itself as we spread throughout the stars," Boberto began.

"The question of what it is to be human became something of absolute importance during the Gene Wars," Cindellou added. "Following the wars, as we clawed our way out of the Cataclysm, we came together and agreed on an answer to that question. Until that time there had been disagreement between biologists about what the definition of a species was. Since then, we've used the same definition. For a population to be human, they must first be interfertile with other human populations. Second, they must be attracted and attractive to other populations."

137

"As the Diaspora began, we had a definition of what it meant to be human, but we had not yet established a way to check," Boberto took up the narration. "Almost fifty years after the establishment of the Ehrewemos colony, The First Expo was sponsored by the Johnson's Fertility Initiative. That Expo not only allowed us to determine whether a population remained human but provided a sense of security and unity that had not been felt since before the Gene Wars.

"All of the colonies and Earth agreed that The Expo was an effective way to check on a population's membership in humanity, but this brought up the question of how often this needed to be checked.

"After much discussion, it was decided that The Expo would be held once a decade. To be considered human, a population can't go a century without meeting the requirements of The Expo. There are two parts: the interfertility testing, which is nicknamed the Baby Pageant, and the attraction and attractiveness portions which together are the Beauty Pageant.

"Since then, The Johnson's Fertility Initiative has not only helped run the interfertility, or Baby Pageant, portion of The Expo, but has also been working to help overcome Cataclysm-induced infertility. Because of their research, and that of other organizations, Earth has once again reached the point where over half of conceptions are natural," Cindellou said as she leaned forward, allowing the camera to see her seriousness.

"Their mission of helping couples conceive perfectly places them in a position to determine what populations are interfertile with each other through not just genetic testing, but through actual production of embryos," Boberto said.

"These embryos also bring the joy of children to otherwise childless couples," Cindellou added in an exuberant tone. "I myself was conceived from interfertility testing for the one hundred and ninety—"

"Now, Cindellou," Boberto said, cutting her off. "Don't tell us what Expo it was. You'll let your age out, and I don't want to believe that you're out of your first century yet."

"Oh, Boberto," Cindellou said slapping him playfully on the arm. "You're just trying to flatter a girl!"

Cindellou's grin faded back to the normal announcer smile

and she continued, "There is another benefit to helping couples with the embryos created during the Baby Pageant. This regular donation of genetic material from the colonies back to Earth allows for the home planet to remain a true mix of all populations and serve as a default genetic blend."

Boberto took up the narration. "I was speaking to one of our guest experts about this for a segment we will be showing in just a few minutes. It will help explain some of the science that goes on in the background of the Baby Pageant. Because the major biomedical centers are still here on Earth, this genetic blend is important to researchers. It makes modifying their research easier if that is needed by the colonies, in case there are differences between Earth and their populations." Boberto paused and gave a self-deprecating grin. "I have to admit he was very excited about the subject, but I was only able to follow the basics of it."

"Well, we will get to view that segment right after these messages from some of The Expo sponsors," Cindellou said, turning to face directly into the camera. "Stay tuned!"

Chuckling as the screen faded back to the presenters, Cindellou said, "I see what you mean about not following everything in his explanation, Boberto. You looked a little lost a few times."

"Well, I have to admit I was. It would have been far worse if I hadn't had the assistance of our staff science reporter," Boberto said grinning wryly. "With that I would like to not only give thanks to our guest, Master Researcher Wohlrabben, but also to Francolin for his help with understanding Herr Wohlrabben."

"Now, we've covered what is involved in the Baby Pageant, and why it is important," Boberto said.

"In more detail than some of us would wish," Cindellou said, as an aside, without looking at Boberto.

Acting as if he hadn't been interrupted, Boberto continued, "As Herr Wohlrabben mentioned in his explanation, detailed information on the interfertility of the different populations won't be available for public release for two decades due to the privacy protections for minors and health-related data."

"Because of this we won't be going into detailed results of the Baby Pageant, but only announcing the overall results. That said—" Cindellou picked up an envelope from the small table between her and Boberto, "we can now announce the results of the first phase of The Expo."

Tearing open the envelope Cindellou pulled out a card and read what was on it before passing it to her co-host to read.

"It is with great excitement that I am able to announce that all populations have met the requirements to complete this decade's Baby Pageant!" Boberto exclaimed, looking into the camera.

"This is great news! Especially for the colony of Crismithian!"

"Their delegation must be ecstatic at this news," Boberto agreed.

"In the last nine Expos they have successfully completed the Beauty Pageant but were just below the cutoff requirements for the Baby Pageant," Cindellou expounded. "This was believed to be due to a nasty outbreak of Spanglish flu on Chrismation that resulted in widespread sterilization. The colony has been aggressively recruiting immigrants to help with their genetic diversity, and it looks like it's worked. If they hadn't completed the Baby Pageant this decade, then they would have not only been determined to no longer be human, but they would have been the first population to have that happen!"

"Based on their past results for the Beauty Pageant, they should have no problems meeting the minimum requirements to complete all of The Expo," Boberto added. "This is exciting news for their colony."

"Well, Boberto, we have our correspondent Trischah Takanai on site with the Chrismithian delegation," Cindellou commented. "She says that it is a veritable party there."

"Sounds like a good time! Why don't we take a look at the festivities," Boberto answered.

Smiling and nodding in agreement, Cindellou turned to face the camera and said, "Trischah, over to you!"

"Welcome back, viewers," Boberto said, grinning into the camera.

"So far we've spent our time talking about the Baby Pageant requirement of The Expo," Cindellou chimed in, "but we haven't gone into detail about the attraction and attractiveness, or Beauty Pageant requirements. For our returning viewers what we are about to cover will be a bit of a review of our previous coverage, but don't worry, we have some interesting segments that will make for more than just a recap!"

"That's right, Cindellou," Boberto said mirthfully. "At the very least, our viewers should get a laugh out of my interview with another scientist!"

"I know I'm looking forward to watching that," Cindellou said, grinning at her co-host.

Flashing a slightly embarrassed smile in response, Boberto continued. "If you have watched coverage of The Expo before, then you will know that the first stage of the Beauty Pageant, or as many people refer to it, 'the swimsuit competition,' has been completed."

"Why do they call it the swimsuit competition, Boberto?" Cindellou asked on cue.

"Good question, Cindellou," Boberto responded. "Historically, pageants were made up of differing parts, but three of the most common were the swimsuit competition, the evening wear competition, and the interview. In the early days of The Expo, the three phases of the Beauty Pageant were jokingly referred to by these names. Since then it's become tradition."

"Swimsuits aren't part of the competition, then?" Cindellou asked.

"Nothing so exciting for us, Cindellou," Boberto answered with a chuckle. "The swimsuit phase in old beauty pageants was all about physical attractiveness, just like the first part of the attractive and attractiveness portion of The Expo."

"I know this phase uses both an MRI and an EEG machine," Cindellou joined in. "I don't understand how these are supposed to determine if someone is attractive or not. Isn't beauty in the retina of the viewer?"

"I had the same questions, Cindellou," Boberto answered. "It turns out the machines don't determine if someone is attractive."

"But, Boberto..." Cindellou started, then trailed off as she looked at her co-host with an expression of doubt.

"They can't be used to determine if someone is attractive, but they can be used to see if *you* think someone else is attractive," Boberto answered her unspoken question.

"That makes more sense and is in line with what the experts told me when we talked about the attraction and attractiveness part of The Expo," Cindellou said. "I know we use EEG in the later parts, so how do they use these machines in *this* part?"

"First let me explain how the test is performed," Boberto said.

"Go ahead."

"Expo participants are connected to the EEG machine. They then lie down inside the MRI machine. Images of other participants are projected above them and all they have to do is look at these images," Boberto explained. "We then find out which other participants they do and do not find attractive."

"That sounds a little too simple to be true," Cindellou said in a tone of doubt.

"Well, it's easy for the participant, but not for our scientists," Boberto replied. "When a participant sees an image of another participant, their brain automatically reacts to them. Our scientists are looking at this activity and decoding whether they feel a sense of attraction or not."

"They can read the participant's thoughts?" Cindellou asked in disbelief.

"They can't decode thoughts, but they can decode what someone is feeling," Boberto answered. "We just need to see if someone feels attraction toward another. This gives us all of the information we need."

"Now hold on, Boberto," Cindellou interjected. "When I was talking with the specialists about the remaining parts of the Beauty Pageant, they were talking about using EEG to determine whether someone feels attraction toward another. If we could do that with just an EEG, then why do we need an MRI machine for the swimsuit competition, but not the rest?"

"You're right, Cindellou, we only use EEG after this in the remainder of the Beauty Pageant," Boberto agreed. "The way it was explained to me is that the MRI machine, using a special scan method called fMRI, has the ability to see more detail. Our scientists use this to calibrate the EEG for us to use after the swimsuit competition is over."

"Hmm," Cindellou responded doubtfully. "I think it might be time to see your interview with Master Researcher Emmlelee Emilsdotter for a more detailed explanation."

"I think you're right, Cindellou," Boberto agreed, turning to face the camera. "We'll be back with that interview right after these messages from our sponsors. Following the interview, the results of the swimsuit competition will be shown."

✧   ✧   ✧

"Well, Boberto, I think Master Researcher Emilsdotter explained things a bit more clearly than you did," Cindellou said with a teasing note in her voice.

"Oh, I definitely agree," Boberto said chuckling.

"And for those of you interested in seeing the results of the swimsuit competition again...," Cindellou added, "it is available on our netsite along with expert commentary and analysis of the data."

"Now that we've caught up to where we are in The Expo," Boberto said, changing the subject, "you were mentioning earlier that you had talked to specialists about the remaining parts of the Beauty Pageant."

"That's right, Boberto. Unlike the previous sections of The Expo, these two are run simultaneously. Both the evening wear competition and the interview phase require interaction between participants from different populations," Cindellou explained.

"This is the first time that many of these participants are going to have any interaction with people from other colonies, isn't that right?" Boberto asked.

"They haven't been deliberately separated, but for many this is the first trip to the home planet for most of them. They've been busy with The Expo, as well as taking time to explore Earth," Cindellou replied. "Our contestants haven't had much time for interaction with different colonial populations until now."

"So, earlier I explained how the swimsuit competition got its name. Would you happen to know how the evening wear competition and interview phase got theirs?" Boberto inquired.

"In historical pageants, the evening wear competition used to be where contestants competed in formal dress for an evening social event. This part of the Beauty Pageant is based on an ancient evening event called 'speed dating,'" Cindellou answered.

"Speed dating? I'd think that even our ancestors wouldn't want to rush through a date with their significant others," Boberto said with a faint tone of disbelief.

"Apparently speed dating was where you would have a group of people that you talked to one at a time for a few minutes," Cindellou explained. "The meeting itself wasn't a date, but a chance to see if you wanted to go on a date with any of the other people."

"Odd name, but I see how it compares to the current evening wear competition," Boberto said musingly.

"Just like the ancient practice, all of our participants dress in their population's evening wear and then meet and talk with the participants from other populations. Now obviously there are too many people to meet all in one evening," Cindellou said with a grin.

"That would be a marathon of a night," Boberto replied with an answering burble.

"For this competition, we take the participants from two populations and pair them for an evening," Cindellou continued. "All of the populations will rotate so that they will have been paired with all of the other populations. The populations will alternate between participating in the evening wear competition and the interview phase, with, of course, some nights off."

"Now that you've told us about this competition, can you tell us more about the interview phase? My understanding is that it, too, is based on an ancient practice," Boberto said.

"The interview phase is based on an ancient idea called a Turing test," Cindellou agreed. "According to historical records, it was originally a test using a pair of computers to try to determine if someone was telling the truth. Two people would talk to each other using only the computer terminal to communicate. Historians today aren't totally sure how it was supposed to work, but we know that it only allows a person to be judged on what they say without any influence from looks, tone, body language, or any other conscious or subconscious signals that could be used to communicate."

"I'd think that we would not want to restrict any communication between people," Boberto said in apparent objection.

"Well, Boberto, we do, and we don't," Cindellou hesitated. "We want the people to communicate, but we want to try to limit miscommunication. Body language and other signals can strongly be affected by different cultures. It's not just the genetics of the different populations that can change."

"I think I understand," Boberto said with a look of apparent dawning comprehension. "We don't want someone to make a gesture that might be friendly in their culture, but—"

"Be something rude in the other person's," Cindellou finished for him.

"So, the swimsuit competition is on looks, the interview phase is on communication with limited cultural influences, and the evening wear competition is a combination of looks, communication, and cultural interaction," Boberto summed up.

"That's the intent," Cindellou agreed.

"Now the one part I am missing is how the EEG comes into play," Boberto said. "We covered how the MRI is used to detect attraction and to calibrate the EEG for the swimsuit competition, but not how the EEG is used."

"Well, Boberto, that calibration of the EEG using the MRI is the key. The EEG lets us detect the brain signals that corresponded to the detection of attraction by the MRI," Cindellou explained.

"So, just like with the MRI..." Boberto said with gradual understanding.

"The participants don't have to report attraction," Cindellou finished. "The EEG alerts our technicians whether attraction is present and lets us score the level of attraction."

"How the scoring works sounds like something we are going to have an expert explain," Boberto said with a smile.

"That's right," Cindellou responded. "It's my turn to show an interview with a scientist. Hopefully you'll find it as informative, and as amusing, as I found your interviews."

"Well, before we go to that interview I just want to let our viewers know what is coming up later in the program," Boberto said, turning to look into the camera. "After Cindellou's interview with Master Researcher Cathrein Smoith, we will come back and have *live* coverage of the delegates from the colonies of Urbanek's Folly and Ehrewemos in the evening wear competition, and in the interview phase we'll visit the delegates from the colonies of Crayona Carpasinus and Castor Nova."

"And we'd like to remind our viewers to vote for their favorite delegates from these events," Cindellou added. "Viewer favorites will be brought in studio for exclusive interviews. If you'd like the delegates to answer your questions, then you can go onto our netsite to submit them. Selected questions will be..."

"Honey..."

The man watching the program turned as the seductive voice called from behind him.

"The kids are asleep. Come to bed." As he looked over his shoulder at his wife, she gave a wink with her nictitating membrane and wiggled her tentacles at him in a little wave before slipping down the hallway out of sight.

"Coming dear," he called excitedly, but quiet enough not to risk waking the children. Hopping up from the couch, he reached out for the remote and mashed the power button with his thumb to turn off the entertainment system before hurrying to bed.

Acknowledgement: The author and editors thank Sandra L.H. Medlock for editorial assistance on this story.

Homo Stellaris —
Working Track Report from the
Tennessee Valley Interstellar Workshop

Robert E. Hampson and Les Johnson

At the Tennessee Valley Interstellar Workshop (TVIW) 2016 Symposium, the Homo Stellaris (People of the Stars) Working Track discussed adaptations that humans and society may undergo to sustain the dream of going to the stars. Participants were divided into groups to discuss subtopics including physiological, sociological, psychological, and political adaptations to support interstellar exploration and colonization. A Synergy Group was formed to synthesize the results and prepare a report. The first conclusion of the Synergy Group was that many of the adaptations would be mission-specific. In particular, exploratory missions that do not result in colonization would have different adaptations than those establishing colonies. Thus, exploratory crews would have to concentrate on adapting humans to space conditions: low gravity, low atmospheric pressure, confinement, isolation, small crew military or mission-oriented social structure, and well-defined mission objectives. In contrast, colonists would need to adapt to a specific planetary environment, community-based social structure, a growing population, and high flexibility in tasks and goals supporting colony growth.

More important, however, was the conclusion that human engineering should be naturally and organically evolved rather than imposed externally. Thus, rather than imposing these changes on a group of interstellar explorers and colonists, humanity as a whole needs to *practice* these changes via a vibrant, self-sustaining space culture with a multigenerational presence at least out to the orbit of Jupiter before serious interstellar missions designs can be contemplated. This space culture should be both for purposes of inevitable technological advancement, and to allow for social structures intrinsic to off-Earth permanent habitation to evolve on their own.

In other words: One means of ensuring that humans adapt to space is to go and live there!

LOCAL PROVING GROUND

One of the assumptions of the Synergy Group was that physiological issues of long-duration space missions would already be solved prior to launching an interstellar mission. However, that assumption also implies that psychological and sociological adaptations to long-duration existence in space have also evolved naturally along with many of the technological precursors for the mission. Unfortunately, this is by no means guaranteed; therefore, the human race must be prepared for failures—of habitat, of health, and of isolated social groups—along the way.

On the other hand, multigenerational isolated community missions (i.e. colonization) represent a social engineering challenge that cannot be adequately duplicated by a strictly solar system–based civilization. The Homo Stellaris group was charged with projecting the factors necessary for anticipated missions one hundred years from now. However, the Synergy Group felt that even one hundred years (assuming that we could launch a space-based society *today*) was not sufficient time to *prove* that any reasonable social framework based on current political models would be viable for a colony totally isolated from Earth. In other words, even with a vibrant solar system society, experimenting with different colony proofs of concept (e.g. asteroid habitats, space stations, or nomadic fleets of ships) can only be partially examined.

Consequently, the social structure of an interstellar colony constitutes a major mission risk even assuming the challenges of propulsion and life extension have already been solved. To offset this risk, small-crew exploratory missions would be more desirable to reduce cost and consumables and allow multiple simultaneous missions, even to the extent of allowing the crew to procreate once the initial mission objectives are completed. The added benefit would be that a small colony "seed" would have a positive effect on the crew as well as the home (Earth) population of potential colonists.

WORLDSHIPS AND GENERATION SHIPS

Readers will note that the above discussion omits multigenerational ships (i.e. slow interstellar craft in which the crew is renewed via procreation during the transit) and "worldships" (hollowed-out asteroids turned into sealed colonies, with a space drive attached). There were two reasons for this omission. The first is that there was a separate working track at TVIW 2016 specifically charged with developing ideas for worldships. The second, and in this context more relevant, reason is that the participants of the Homo Stellaris working track felt that the very concept of worldships was antithetical to interstellar colonization. It may likely be the case that a natural component of solar system colonization will be to build structures (or hollowed-out asteroids) which contain complete biospheres. Such a closed-loop life system will be self-sustaining and can support humans as a compromise between wholly planet-based lifestyles versus ship or station-based lifestyles. A worldship may serve as a larger, more robust "space station" or may constitute a stand-alone colony in its own right.

Worldships are an end in and of themselves. Moving such a large biosphere to another star system would likely take centuries. If a worldship would be viable for the projected duration of the mission, then it would most likely be viable well in excess of that timeline. Thus, a worldship *is* a colony; once established, attaching engines or even an interstellar drive to a worldship may provide mobility, but to what end? Furthermore, if it is used merely as a vessel to transport colony and crew, then what is the guarantee that they will want to leave the habitat once the destination is reached? Certainly a worldship can be designed to last only for the duration of transit, but that requires several dangerous assumptions; mainly, that the design life is accurate and critical systems will not fail prior to reaching the destination. Modern engineering is not so perfect that humanity can guarantee zero defects prior to the planned date of obsolescence.

However, worldships are not just transport vessels. They *are* important colony *practice* environments. The more experience we gain from engineering biospheres and ecosystems (not to mention self-contained communities), the better prepared we will be for dealing with new planetary environments.

ADAPTING HUMANS

Having rejected worldships as colony transport, the Synergy Group also felt that relying on multigenerational transport was also less desirable than ensuring that humans could make an interstellar journey in a single lifetime. Thus, emphasis should be placed on gene selection that would extend lifespan, provide greater intrinsic biological space-radiation resistance, and optimize the human body for a lower gravity regimen. A robust space-based society will likely already seek these genetic modifications through their multiple-generation communities off Earth. This also points back to the advantage of allowing space communities within the solar system to be fundamental breeding and proving grounds for the crew composition and colony population.

In addition to intrinsic physical life extension and robustness in deep space, interstellar exploration will most likely require some form of metabolic suspension. While such medical technology is still science fiction, it has its roots in present-day advances in surgical techniques, in the as-yet-unexplored functions resident in what has been called junk DNA, and in lessons learned from vertebrate animals which can successfully survive freezing temperatures without damage to cells caused by the formation of ice crystals. An optimal scenario might be one in which a crew splits into shifts such that some would be in metabolic suspension at any given time. Cycling crew in and out of hibernation would allow for sufficient crew on-watch to deal with both routine and emergency situations at any point in the mission. The number and composition of these shifts will rely heavily on lessons learned from submarine crews, space habitat simulations (such as Antarctic bases and MARS500), and practice in the form of future solar system–based habitats.

Between increased lifespan, and deferred per-person mission activity equal to only twenty-five to fifty percent of the total transit time, mission durations greater than fifty years would

reduce the subjective passage of time to scenarios with compara-
tively low social-engineering requirements compared to generation
ships or worldships. As an added advantage, the percentage of
the spacecraft and colonization materials devoted to life support
might be reduced accordingly. To further extend those resources,
the Synergy Group suggested that smaller-scale automated cargo
probes could be launched in advance, enabling rendezvous and
resupply at key navigational waypoints for the crewed mission.
These probes would constitute additional proof of concept for
mission engineering as well as progressively more advanced survey
and reconnaissance of target systems.

MISSION TARGETS

Finally, the working track was tasked with describing an interstellar mission based on their discussions. Mission parameters were
to assume conventional space drives based on known physics, the
necessary precursors for a space-based society, and the societal
will to undertake such a goal. The working track recommended
selection of initial target systems at the shortest possible range,
that is, not beyond eighteen to twenty light-years. Assuming about
a hundred years to develop the technology for continuous thrust
(allowing final velocities approaching single-digit percentages—
one to nine percent—of light-speed), the newly discovered rocky
planet at Proxima Centauri b is only about forty to fifty years
away. More likely candidate worlds for exploration such as Wolf
1061, Gliese 876, Gliese 682, and Gliese 832 would be around
one hundred fifty years away. Superior drive performance and
improvements in life-extension would allow more-distant destinations, coupled to a reduction of mission times, and simplification
in mission logistics.

The primary conclusion of the Homo Stellaris working track
was that many of the problems may not be completely *solved*, but
would be greatly aided by establishing a presence off Earth and
throughout the solar system. While many Earth-bound issues
would remain, the individuals who would voluntarily go to space
in the near future would be the same type—and possibly even
the same individuals—who would undertake interstellar missions
in the far future. Thus, simply having a presence throughout
the solar system would serve as the developmental basis for the

adaptations required for one or more interstellar missions. Such a mission may very well be within our reach in the next century as we transition from *Homo terranus*, to *Homo solaris*, and finally, into *Homo stellaris*!

Acknowledgement: The editors thank
Dr. Charles E. Gannon, Sarah and Dan Hoyt,
Connie Trieber, Chris Oakley, Carol Tevepaugh
and Doug Loss for their contributions to the
Homo Stellaris Synergy Group report from TVIW2016.
Philip Wohlrab, Cathe Smith and Sandra Medlock
assisted with moderating and facilitating the
discussions of the Homo Stellaris working track.

Time Flies

Kevin J. Anderson

Kevin J. Anderson is the author of over 145 books, fifty-six of which have appeared on national or international bestseller lists. He has won or been nominated for the Bram Stoker, Shamus, Hugo, Nebula, Scribe, and Colorado Book Awards. He is a noted Futurist who has lectured before many world-class venues and is the director of the Certificate in Publishing for Western State Colorado University. He and his wife have been married for twenty-five years; they live in a castle (yes, a castle!) in the Rocky Mountains of Colorado. For more information, go to: Wordfire.com

A life is measured in seconds, days, years—even centuries, now that we have genetic modifications. When all is said and done, each life has a set number of minutes, but no one knows that number, which is set only by God and by destiny. Those minutes wind down to zero, *tick tock*, over the course of a life, and ultimately, we can't change the number that has been ordained for us. But if we can speed up or slow down our frame of reference, we can make the outside *objective* time last for as long as we like.

That's why starship flyers seem to be immortal from an outsider's point of view. We have the same number of minutes as anyone else, but thanks to our special metabolic modifications, we knew how to shut down our internal clocks for the centuries, or millennia, that we drift between the stars. From our point of view, it's just a regular lifespan, but stretched out like a thin membrane across infinity. We can control our own perception of time, and the ship is like its own separate continuum moving endlessly throughout the galaxy.

As engineer, I was the first member of the crew to speed back up to realtime as the *Time's Arrow* approached the new star system. It's my job to run a check on all systems, though the old starship had been so modified and reinforced that nothing sort of a cosmic disaster would have caused so much as a hiccup. But I don't like to think I'm obsolete.

According to the flight plan, we had just spent two hundred years en route to this average star system with its one cataloged colony planet, called Irrac if the old records were still accurate. Captain Dorothea had chosen it as the next stop on our ever-wandering trade route. No one among the small crew had disagreed with her, not that we ever did. We didn't have any place else to go. *Time's Arrow* was our own little traveling universe, and everything else was just a side trip.

Moving through the silent ship with the other nine crewmembers still in slowtime, I checked the engine systems, the fuel, and the water levels, which were supplemented by interstellar gases we had scooped up over the past couple of centuries. Planetfall was still a month away, but we had preparations to make. I saw the others like statues lounging about wherever they had decided to crank down their metabolism to near-zero entropy, like the cozy heater in our rec room turned down to the tiniest pilot light.

I warmed up the rec room, getting it ready for our crew meeting as soon as the others returned to the normal flow of time. Next, I went to the command module, the bridge portion of the ship, which was nearly always empty except for when we reached a planet and the interesting activity started.

In the cold of the command module, I wore a warm sweater. Since I would only be there for a few moments, I saw no need to waste the time and energy to warm up this section of the ship . . . not yet. While everyone else still blissfully and invisibly passed the time without wasting minutes of their lifespan, I transmitted our prerecorded welcome message to the system ahead, announcing to Irrac that the *Time's Arrow* was a private commercial vessel filled with exotic cargo that we wished to trade. Our last stop, two centuries ago in subjective time, was the inhabited moon Jherilla, circling a gas giant. The people on the planet ahead probably knew little of their neighbors anyway, and we didn't want to scare the locals by telling them just how long we'd really been on our voyage.

After sending my introductions, I asked Irrac to send a transmission burst with samplers of their language and dialect, information on their culture and customs, so the *Time's Arrow* could more easily interact when we arrived. Since all planets were so isolated and scattered, no one could guess what sort of welcome we might receive, and it was good to be prepared. We had defensive weapons, but rarely needed to use them. No one on Irrac was expecting us, and we hoped for a nice, profitable landing.

I decided to allow a week before checking again to see their response.

Still moving at normal speed, I rejoined the crew where they all sat motionless in the galley. Everyone wore comfortable uniforms; no need to be formal after all this time. Captain Dorothea rested in her favorite chair, beautiful, in her mid-forties, with smooth features, a pointed chin, a sharp gaze when she was displeased and a sparkle when she was happy. Right now, her eyes just looked like glass. With her near-zero metabolism, Dorothea couldn't see anything moving at my speed. I gave her a quick peck on the cheek, which was much less interesting than when she actually participated in a kiss. I chose an empty chair next to her at the small dining table and dropped back down into slowtime.

Long ago when the *Time's Arrow* first set off on its endless voyage, the crew would all go to quarters or settle in special travel beds for the long journey ahead, knowing we wouldn't move an inch for centuries at a time. Eventually we realized that we noticed nothing in slowtime anyway, so we no longer bothered with the formalities.

Time's Arrow was a privately owned, self-sufficient commercial ship, and we had all joined aboard for the adventure, a life of trading among the colonies scattered across the galaxy, and that's exactly what we had done. When we departed from Earth ten thousand years ago, objective time, none of us really grasped the timescales involved...

After waiting a week, which passed faster than the blink of an eye from my point of view, I sped up to realtime again and returned to the command module to check for a response from

Irrac. Now that the ship was closer, scans showed a viable planet ahead. I was glad to see that the colonists had responded with friendly surprise, transmitting complete files about their environment, culture, history, as well as their eagerness to welcome us. If the signs didn't look right, we could have opted to bypass the system and head off to the next destination, but Irrac seemed fine, so I allowed the ship to continue its three-week journey to the planet.

Dorothea sped up first, then our eight other crewmembers who served as scouts, botanists, manual labor, and "other duties as assigned." Other than Doctor Max, their roles were ambiguous, not that it mattered. Everyone did his or her own chores aboard ship, and the ship did most everything else.

Delman accepted the duty as chef and prepared our wakeup meal, a stew of preserved meats and vegetables we had picked up at one of our other stops, I couldn't remember which. Most of us weren't even hungry, because in our timeframe it hadn't been very long since we'd eaten our traditional bon voyage meal after leaving Jherilla. Delman was a good cook, though, and he made excellent use of the various flavorings and spices available. I filled my plate and sat next to Captain Dorothea, presenting my report as we all ate. No need to waste minutes on a boring formal meeting when we could take care of the details at dinner conversation.

"The people on Irrac seem friendly, thrilled to receive out-of-town visitors." I played some of the recordings from the colony leader. The evolution of dialect and language over the centuries made him difficult to understand, but thanks to the records transmitted by the colonists, we were easily able to update our language chips. I displayed images of Irrac's grassy hills, sweeping green fields, exotic animals, thick forests, pleasant looking cities and villages.

"Beautiful place," said Amos, one of our scouts. "A lot better than most of the hardscrabble outposts we see."

"Looks like they hit the jackpot in the colony roulette," Dorothea said. "Are they covering anything up, Garrett?"

"Nothing major," I said. I had only scanned through the records once, but the signs would be obvious. We'd encountered nightmare settlements before.

After humanity spread across thousands of colony worlds,

societies changed and evolved in unpredictable ways over centuries of isolation. We'd gotten good at detecting the signs of horrific dictatorships or repressive religious societies. Here, though, the bright colors and casual nature of the Irrac garments, the openness of their architecture, how freely the people moved about their streets, their free conversation and interactions all implied a normal healthy society.

"I sent them our detailed manifest from the cargo hold," I said. "They probably don't understand what half of the stuff is, but we'll show it off anyway. Right now, they're scrambling to find goods or raw materials that we'll accept in trade. It looks like a good place to stop, Captain."

Doctor Max agreed with his characteristic chuckle, and the others looked forward to another shore leave, even though they had barely noticed the two centuries of passage since the last one on Jherilla. Aboard the *Time's Arrow*, it was just one planet after another with a fast-forward journey in between.

But Amos was surly. He had grown clearly impatient at the previous three or four stops. "I'm sick of this routine. I really can't stand the shit anymore."

"Then stay awake in realtime and redecorate your quarters," Dorothea said. "Take up a hobby." Over the past couple awakenings, she'd grown impatient with his attitude.

"It just feels pointless." Amos rested his elbows on the table, looking down at his barely eaten food. "We had a great haul at Jherilla. We're all rich, but what do we do with it?"

"We're going to trade it for other things on Irrac, and then we'll do it again. And again and again," said Elber, one of the other scouts. "It's what you signed up for."

"Ten thousand years ago." He picked at his vegetables and meat.

"I thought you wanted to see the galaxy," Dorothea said.

"I've seen it." Amos glanced at the images of Irrac, and his face took on a wistful look. "Maybe I just want to settle down."

"Go right ahead," said Delman. "Plant your roots in the dirt. More shares for the rest of us."

Frustrated, Amos scowled at him. "Shares for what? What is it for?"

I scooped the last few bites off my plate and stood. I'd never been much interested in conversation, and especially not arguments. "I've got to prep for orbit and landing. When I get the engines reset and the orbital curves plotted, I'm dropping back down to slowtime. You all can debate the meaning of life for the next three weeks if you like. I don't want to waste my time on something I can't answer."

I left the galley and returned to the command module and transmitted a positive response to the colony, this time using perfect Irrac dialect. And since I'd have to be here when we reached the planet so I could guide the ship in, I didn't even bother to leave the bridge deck. I took a seat in the padded chair and dropped down to zero.

When I sped back up to realtime three weeks later, the planet filled my view as if a huge world had popped out right in front of me. Irrac had all the right colors: blue, green, and brown with swirls of white clouds, breathable atmosphere, temperate climate. The scan showed more than a hundred cities scattered across several continents. The *Time's Arrow* comm log showed countless unanswered messages and reports the colonists had transmitted to us as they grew excited at our approach, although I had made it abundantly clear in the last transmission that the crew would be in slowtime and no one could respond.

The starship had automatically gone into orbit. Dorothea came to the bridge and placed a hand on my shoulder as we both observed the planet. "It all looks good," I said. "No military buildup, no panic or unrest."

"Then take us down. No need to be skittish. We spent two centuries getting here, and you and I only woke once at the same time for a little exercise." She smiled.

I put my hand on top of hers. "That was a couple of hours well spent." Yes, even after thousands of years she was still a damn fine woman.

We were surprised to discover that Amos had returned to realtime a full day before the rest of us so he could review the Irrac transmissions in private. He'd even used the comm to speak to people down in the main city, even though such an unauthorized transmission was technically forbidden.

The other seven crewmembers sped up on their own schedule, and everyone was fully awake and alert as *Time's Arrow* descended to a large open area outside the main city. It wasn't an actual spaceport, but our ship easily made do.

A large welcoming committee had gathered for us, cheering crowds waving colorful banners to welcome the exotic visitors from so far away. I thought of parades and marching bands on Earth, but I wondered if anyone other than the *Time's Arrow* crew even remembered such things from more than ten thousand years in the past.

Though I'd been the one to make initial contact with the colony leader, Dorothea was the face of our expedition. I prefer to stay in the background, not needing the applause or the attention; in fact, I don't like it. As we emerged, I stood with my companions in formal uniforms, just another part of the crew. Captain Dorothea gave an engaging and inspirational speech about our journey, told of the wonders *Time's Arrow* had seen and the exotic goods we brought. She explained how much she appreciated such a warm welcome from the people of Irrac.

I had heard the speech close to thirty times before, and so, without anyone noticing, I slowed down my internal clock so that the boring part sped by in just a few subjective seconds.

Oddly enough, Amos was singled out and welcomed by the people. All of Irrac knew who he was because of the direct conversations he'd sent from the ship while the rest of us were in slowtime. Families came to meet Amos, starry-eyed, giving him warm embraces. He surprised us further when he informed the captain he'd been specially invited to join a large clan grouping for feasts and parties. Normally, the crew kept close to one another, but shore leave was shore leave. When the opportunity arose, we all caroused and sampled the local pleasures. This wasn't overly unusual.

For the next several days we unloaded the extensive goods we had carried from planet to planet. The exotic items, the metals and materials, strange artwork, music recordings, jewelry, incredible food samples from other worlds—to the humble people of Irrac it was all breathtakingly priceless. The town leaders despaired that they had nothing we would find of value in trade.

What they didn't understand was that a normal, pleasant colony was a rare thing, and their forested hills provided a wealth of beautiful lumber. The wood seemed all too common and of little value to Irrac, but Dorothea insisted it was worth its weight in gold, a cliché that the dialect chips translated into the appropriate local phrasing. Gold, in fact, was easily obtained on any number of rocky asteroids, but tall trees were extremely valuable, and the captain knew she could sell the lumber at a premium on some other austere colony where real wood was a rarity. Dorothea and the colony leader were both satisfied with their bargain, and our mission to this world was a complete success.

Except when Amos came up to us on the day before our departure. I had just emerged from the cargo hold, wiping my hands on a towel after storing more than five metric tons of fresh wood, and saw the crewman nervously reporting to the captain. "I've made up my mind, Dorothea. I'm staying here."

A pretty, young woman stood next to Amos, a little blander than his usual type. She had a little girl at her side no more than three years old, clutching her mother's hand. They both stared in awe at our landed starship.

Dorothea's brow drew together, and her eyes didn't have much of a sparkle at the moment. "You're staying behind . . . how?"

"I'm opting out of my contract. I'd like a reasonable share of profits so I can set up a stake and make a good home for myself here. I'd like to live the rest of my life on Irrac. It's the best place we've seen in a long time." He reached out to take the young woman's hand, and she gave him an adoring look. "Alila's family has offered her to me in marriage. She lost her husband last year, and her beautiful little daughter here has no father." He swallowed and said, "I've always wanted a family. Now I can have everything I want. If I stay here."

I stepped forward, still rubbing my hands on the towel. "If you do that, there's no turning back, Amos. You want to spend the rest of your life in realtime?"

"I can always drop down if I need to, turn into Rip Van Winkle." He chuckled at the joke, which he knew that no one understood but us. "Why would I want to? I'll have a home, a wife, a daughter to adopt, maybe other children someday. I'll be comfortable and happy for my lifespan, however long that might be. Isn't that what we all want? I can finally have it."

"Your life will be over in the blink of an eye," Dorothea said. "On the *Time's Arrow*, you'll keep traveling for centuries and centuries."

Amos shook his head. "If I choose to live out my time here with Alila, what's wrong with that?"

"Nothing at all," Dorothea said with a resigned sigh. "Nothing at all. Go with my blessing. I know you've been dissatisfied for a long time. We'll figure out how to pay you in some currency that's valuable to these people. The ship will just get by with one less crew member from now on."

I refrained from pointing out that *Time's Arrow* could get by with almost no crew at all, but the ten of us had been together for so long, this was like losing a family member.

Over the following day, we all said our farewells. Every hour, the condensers had filtered and stored air, so our reserves were at capacity. I loaded up our reservoirs with water from a local lake. We were ready to go.

I kept expecting Amos to lose his nerve and change his mind, but he didn't. The clan units welcomed him, and he seemed anxious to begin his life with his new wife and daughter. As *Time's Arrow* departed, leaving Irrac, I zoomed in on the image of the crowds waving goodbye. I saw Amos staring at us with a triumphant and satisfied expression.

He would be dust by the next time I had a chance to think about him.

Six more centuries passed en route to our next destination, a planet called Riece on the old charts. It was a long trip, but the point of our constant nomadic mission was not to reach any particular place. The journey itself was its own goal, one world after another. The expansive galaxy, all the places we could go and see, and all the history we were unrolling behind us—*that* was what mattered to me.

Time's Arrow was hardened and self-sufficient, with low-level AI monitoring systems to take care of any routine maintenance, so the ship could effectively fly through empty space forever. Even so, I still made it a habit to return to realtime once every year or two, just to have a look around. Yes, I was burning minutes of my lifespan, but I enjoyed the solitude, slowly wandering through the ship, looking at the motionless slowtime figures of my eight remaining fellow crewmembers. Sometimes a person just likes to have time to think.

I would sit alone in the command module and stare out at the empty sea of stars and imagine the distances, how many uncounted years that starlight had been shining, the same way that *Time's Arrow* had been flying through the void. From my perspective, we had left Irrac only a few days ago, but I knew that by now Amos was long dead, as was his young wife. The pretty little daughter would have grown up, started her own family, grown old, and died, along with the next generation and the next.

But we were still here, still flying.

Amos could have still been alive, but he'd chosen to stay behind. I hoped he didn't regret his decision after we left. I tried to imagine myself doing that, but couldn't. Life wasn't a race or an endurance test, and the universe didn't hand out a prize for the person who managed to stick around longest. As I looked out at the stars, I imagined an ancient sea captain gazing across the uncharted ocean and hoping there must be some new land beyond the horizon.

What I wanted was to see the universe, but also to see the universe *change*. With my own eyes, I was watching the cosmos evolve, stars form, nebulas coalesce. It was like observing the universe on fast-forward.

The *Time's Arrow* crew was my family, and Captain Dorothea was my lover. Often on my brief maintenance walks, when I was the only one in realtime, I would look at her calm and beautiful face, the frozen expression reflecting whatever thoughts had been in her mind when she'd dropped down to slowtime.

The captain and I had our arrangement. We would agree on a schedule and both speed up simultaneously so we'd have the silent ship to ourselves, privacy for our lovemaking. I would often accelerate a few minutes before Dorothea awoke so I could be standing there with a smile. When she became animated and looked at me with that sparkle of anticipation in her eyes, I'd fold my arms around her and pull her close. With the rest of the crew unaware, she didn't have to be the captain and I didn't have to be the engineer. We became just two human bodies touching. Captain Dorothea was independent and firm, I liked my alone time, but we both needed a certain amount of companionship. I'm not sure any of the crew even knew about our relationship.

Dorothea and I would sometimes take an hour or more to make love, in no hurry. Afterward, we would just lie together in her captain's bunk, feeling the warmth of skin and sweat. This time, we talked about Amos, because I couldn't stop thinking of him on Irrac, his home, and his family. He had a bed, a kitchen, chores to do, bills to pay, mundane concerns that none of us aboard *Time's Arrow* had thought about in millennia. And he was long gone, though I remembered seeing him only a few hours ago.

"You think he was happy?" Dorothea asked.

"I think he's dead now. Was the happiness worth burning all of his years so fast?"

She mused for a long moment before finally answering, "The measure of a life is what you do with it, not how long it lasts."

"I liked Amos, so I'll be optimistic," I said, stroking her hair. "But I'm not going to turn the ship around and fly back to Irrac to see if his name is still listed in their history books."

She chuckled and kissed me. "We don't look at history books, Garrett. They're not relevant. We're just moving forward, always forward."

We stayed awake long enough to make love again, and then we both dropped to slowtime while still lying in her bunk. I would speed up in a year or two anyway for my maintenance round, and I could get dressed then.

Toward the end of the six-century run to Riece, I grew bored and only sped up every fifteen years or so. Constantly outbound for millennia, by now *Time's Arrow* had reached the far fringe and some of the most isolated colonies in the expansion of the human race. Any colonists who had traveled this far either had truly grandiose dreams or they were running from something.

Over the past several decades on our journey, I had picked up a few scattered broadcast transmissions from Riece, so I knew the colony was still there. When we were a year out from the system, I returned to realtime, intending to do a little preliminary snooping—and I immediately smelled smoke, an old, bitter tang of wrongness that hadn't been fully scrubbed by the near-dormant life-support systems.

Feeling sudden urgency, I accelerated, moving even faster than realtime. In a blur I raced to the monitoring systems, found that there had indeed been a serious fire in the rec room—three years ago, by ship's time. After so long, there was no need to hurry, but I still ran.

The rec room was a scorched mess, smoky, still sealed. The basic fire-suppression systems must have failed. *Time's Arrow* was ten thousand years old, after all. I ran a diagnostic every century, but maybe I had grown complacent. The heater unit, which Captain Dorothea considered a homey touch, had malfunctioned, starting a fire.

Three crew members had burned to death, unrecognizable at first glance. Their bodies were charred. One had collapsed on the floor. I soon identified them as Delman, Thea, and Olivia. They had been sitting around the table, and I saw playing cards. A poker game. Apparently, they had arranged to speed up to realtime to enjoy a card game while the rest of us were dormant. Neither I nor Captain Dorothea had been invited. I wondered how long that had been going on.

For some reason, the three had dropped into slowtime after finishing a round, maybe taking a break, and they had burned to death before even realizing what was going on. Olivia, the one collapsed on the floor, had apparently sped up to normal... but too late. She had dropped from the flames or smoke inhalation. All three were truly dead, not just slowed down to zero.

Worse, although the main fire in the rec room had been suppressed by the automated systems, an electrical fire had raced through the conduits, causing more severe damage to other parts of the ship.

I ran up to the bridge to scan the monitors and see how bad the situation was. The command module was intact—it was the most secure bastion on the vessel and could fly as its own self-contained starship—but one of the stern engines and some systems in the cargo bay were severely damaged.

I checked the log and navigation systems, verifying that we were still a year out from Riece. Even if I spent weeks, or even months, of my own time, alone, I couldn't complete the repairs here. I could patch up some of the basic systems but I needed a place to land, a technological society, or at least the raw materials for the fabricators. I transmitted a distress signal to the distant colony planet, letting them know we were coming. This would give them months more notice than we usually did, but *Time's Arrow* needed their help.

I scanned for more transmissions from Riece, but the people were quiet—either shy, reclusive, or they had fallen back to more primitive technology, which might make full repairs more challenging for me.

Keeping the news to myself for now—what could the captain

or anyone else do at this point?—I stayed in realtime for a week inspecting ship systems, doing the grim work myself. I took the three bodies to the infirmary and packaged them up. Doctor Max might want to have a look, but I doubted he'd have the ambition to perform an autopsy. There was nothing he could do for Delman, Thea, and Olivia.

I fixed the systems I could, but the burn was especially extensive in the cargo-bay controls. Finally, I roused Captain Dorothea from slowtime so I could give my full report. When I told her, I watched the hard beauty of her face melt into grief. She realized the loss of our friends, our family, how greatly our crew was diminished, not only after Amos left us at Irrac, but now three dead in the fire. We only had six crewmembers remaining.

Since the two of us were still alone in our own time continuum, I held her, sharing strength. When she composed herself, she asked, "We're still on course to Riece? You're confident we can make repairs there?"

I shrugged. "Reasonably confident. I can't get a good read on their tech level, but I can make do with whatever they've got."

"Let's wait until we're a month out. We can tell the others when they awaken. We'll all go down to Riece and make everything right."

More sluggish than usual due to the damage in the cargo module's stern engine, *Time's Arrow* headed down toward the surface of Riece. I still didn't know what to expect.

We had entered orbit, identified the largest cities, verified that Riece was a technological civilization, although one with minimal radiofrequency chatter. I had contacted them from the command module, describing our situation and listing the cargo we had to trade. Riece sent only intermittent acknowledgements from a bland-looking shorthaired woman, Renna Qo, who claimed to speak for the people. She gave coordinates for us to land, showing neither threat nor exuberance. In order to repair the lower cargo modules and the engine, we didn't have much choice.

"You'd think they'd be a little more excited to receive out-of-town visitors," Dorothea muttered.

The four other crewmembers were still dealing with the loss of Delman, Thea, and Olivia in the rec room fire. As engineer, I explained that the damage was mostly confined to the lower module of the ship and that *Time's Arrow* could fly just fine for now even with the reduced crew.

"The main thing is to make repairs," I said. "I've already taken care of the electronics and life support. The command module and the living quarters are fine. Our fabricators can make the necessary components from basic planetary materials even if Riece doesn't have the manufacturing capability. Given a stable work environment, I'll fix the cargo module and the supplementary stern engine in a few days."

"Sounds like a week of vacation on Riece. Let's hope it's a nice place," said Doctor Max.

"With pretty girls at least," said Elber. "Or pretty guys." The other two remaining scouts, Henrik and Anya, agreed.

I set us down where Renna Qo had directed in a textbook maneuver, despite the sluggish response from the lower half of the ship. Only about fifty people had gathered on the fringe of the landing area, though I had expected a much larger crowd. The Riece natives were gaunt, pale skinned, and wore drab clothes. They had an odd sameness about them, and they didn't even seem to talk with one another. We had scanned for weapons and found none. The people weren't a threat, just uninterested.

I remembered the banners and the cheering reception on Irrac. This was quiet and subdued.

While Captain Dorothea, Doctor Max, and the three scouts emerged for the traditional raised hand and "I come in peace" greeting, I went to the cargo vault to prepare our wares for display. Even after all this time, the air inside the cargo hold still held the sweet resinous scent of the sealed and preserved lumber. Other crates of goods were stacked up and neatly inventoried after our numerous stops along the way.

Now that I was down in the cargo bay myself, I looked around. I hadn't been fully candid with the crew about just how damaged the lower-level systems were. Yet, after ten thousand years of voyaging, every system could use an overhaul. I hoped Riece had the right facilities to let me accomplish my work.

I thought enviously of Amos staying behind on Irrac. That pleasant world would have been a much nicer place to work, while Riece seemed flat and exhausted, compatible with human life but not an Eden by any means.

I opened the cargo-bay doors so the gathered natives could see our warehouse. At the nearby disembarkation ramp, the captain, Doctor Max, Elber, Anya, and Henrik came forward to greet the people. The Riece reception committee stood eerily silent without cheers or whistles, not even background conversation. I spotted Renna Qo, the inflectionless woman who had served as their spokesperson. She stepped forward.

Dorothea acknowledged her. "Thank you for receiving the *Time's Arrow*. Our journey from planet to planet has lasted thousands of years, and now we need your help. We bring many goods to trade, along with the full database of our travels so you can learn what humanity has been doing over the millennia since you left home."

She paused, waiting for a response. The people of Riece stepped forward as if to greet them. Renna Qo's expression remained flat.

Dorothea continued, "Your planet is very isolated, and I know you haven't received many offworld visitors. We look forward to learning more about you." She extended a hand as Renna came close. "We'll also need a place to make repairs to our ship. We'll be self-sufficient, though we'd like your support."

"We cannot allow the contamination," said Renna Qo. "The mind of Riece has decided. You will be purged."

Immediately wary, Elber, Henrik, and Anya drew their weapons. Doctor Max raised his hands. "Now, now! There's no reason to—"

The crowd of fifty fell silent, moving in an eerie unison. They concentrated, then hummed.

Dorothea's head exploded.

Even sheltered inside the cargo bay, I felt a pounding roar in my head, destructive thoughts like a telepathic battering ram. Elber screamed and dropped to his knees, clutching his head as if to hold his skull together. Doctor Max yelled, and blood leaked out of his eyes and ears, then his brains boiled out of his eye sockets. Henrik and Anya turned to run, eyes squeezed shut, screaming as they ran back up the ramp, but they managed only three steps before both fell flat as if someone had struck them on the head with a heavy club.

Even shielded inside the cargo hold, I felt the throbbing through the hull. The natives pressed closer as if to storm the ship.

Barely able to see, I staggered to the cargo-bay controls and sealed the door. The heavily reinforced slab slammed shut. The people pressed around the ship, and waves of their deadly thoughts vibrated through the hull. I didn't know how long I could last against that onslaught.

The lower cargo module with its damaged systems and the malfunctioning stern engine would be dead weight in the planet's gravity well. I knew what I had to do. Sooner or later their telepathic battering ram would kill me. I couldn't stick around to find out.

Sweating, seeing red static in my vision, I reached the command module and threw myself into the bridge seat. The pounding in my brain was so great I couldn't think of grief or shock, couldn't react to seeing Dorothea's head split open, to the loss of my remaining comrades. The people of Riece pressed forward against the ship.

The upper section of *Time's Arrow,* the command module and living quarters, had everything I needed. The cargo half of the ship was a hindrance, and I couldn't take off with it. I blasted the explosive bolts and separated the command and habitation module from the bottom section, burning more fuel than was wise, but I had to tear away from the leftover carcass of the ship. I left my dead captain, my crewmates, and all our cargo surrounded by murderous natives.

As the command module rose higher and higher, roaring away from the colony, I felt my thoughts grow still. The aggressive mental attack faded with distance. I touched my face and found blood coming out of my nose.

As soon as I reached orbit, I set course at random, a straight line out of the Riece system. I just needed to get far away, find a place where I could recover.

As soon as the ship was flying, I fled into slowtime so I could escape, all by myself.

When I was a year away from Riece, I returned to realtime in order to assess where I was, what had happened, and what my options were. I had no captain, no doctor, no scouts, no crewmates...no family or friends. *Time's Arrow* was so far out on the rim, I would have to change course and follow along the spiral for centuries before I reached the next known colony. Even so, it was only a name on an old star map that was impossibly out of date. If the people of Riece had evolved into telekinetic, xenophobic murderers, how could I know what the next planet might hold?

And what would I do there when I arrived? I had only the command and habitation module, and no cargo to trade. Yes, I had information and the ship's records of our journey, and many colony worlds would be happy to receive an exotic traveler such as myself. But I still had reservations.

Time's Arrow could travel forever, for all intents and purposes, but entering orbit, dropping down to a planet's surface, and then climbing back up out of the gravity well burned a lot of fuel and added the most wear and tear to the ship.

I thought again of how I hoped Amos had lived a long and happy life on Irrac. I had criticized his decision at the time, but maybe Amos had been right. He was long, long dead, but here I was alone on a ship on the far rim, where colonies were breathtakingly far apart. I could turn around and head back into the more densely populated star systems... or I could keep going outward.

I took stock. Since it was only me aboard, and since the ship's systems were highly efficient, I had enough food, water, and air to supply myself for centuries even in my subjective frame of reference, and that was more than enough. After all my life, after ten thousand objective years flying across the galaxy in *Time's Arrow*, I didn't like the idea of going backward.

In fact, as I sat in the command module and looked at the sparse star field before me, I realized that if I dropped down to my slowest possible subjective speed after accelerating the ship, I could keep cruising practically until the end of time. I would drop back into realtime once every thousand years or so and have a look around.

It seemed as good as my other options.

There's something unique and egotistical about knowing you're the last human—at least this version of human—in all of existence, in all of the universe. Given the length of our journey, maybe I already was.

If I kept flying, I'd have a front-row seat to watch the evolution and maybe even the end of the universe. I had my lifespan, my allotted number of minutes, whatever it was—and I also had the ability to stretch out that lifespan as long as I could.

I had forever, and forever seemed a worthy goal.

I set the course of the *Time's Arrow* out of the galaxy and flew into the great unknown. I triggered myself to awaken once a century, once a millennium, once every ten thousand years.

The ship drifted on, and my minutes ticked away. I wondered how far I would get.

But every time I awakened I saw the same thing. Blackness and stars.

Blackness and stars.

Blackness and stars.

Something no one else had ever seen.

Blackness and stars.

Blackness and stars.

Forever.

Our Worldship Broke!

Jim Beall

Jim Beall (BS-Math, MBA, PE) has been a nuclear engineer for over forty years, a war gamer for over fifty, and an avid reader of science fiction for even longer. His experience in engineering and power systems began as a naval officer after surviving his much-dreaded interview by Hyman G. Rickover, the Lord Admiral and High Priest of nuclear engineering. Subsequent experience includes design, construction, inspection, enforcement, and assessment with a nuclear utility, an architect engineering firm, and the US Nuclear Regulatory Commission (USNRC). Along the way, he either learned or derived the safety-focused mantra of the nuclear engineer, and fully expects that it will be applied most rigorously to worldship design.

"The reason I asked to speak with you here, in this place, is to tell you that something has broken."

Perhaps we are meeting in the heart of the Tabernacle, with you in the vestments of the High Priest and me in the raiment of the ArchDeacon of Engineering. Or, maybe we are on the bridge of our great vessel and you're wearing the glittering dress uniform of a ship captain, with me your engineer. There are countless other possibilities—from business suits to no clothes at all!—but my fear in every case will be the same.

It's not my fault!

"Please don't excommunicate, execute, or recycle me!"

I am not going to try to blame our ancestors. Whether I am

reading from scripture, logs, or reports, I will attempt to convince you that failures have occurred before, and they simply happen no matter what. After all, we have been traveling in the vacuum of space toward our destination star for a very long time.

"Raise not your staff to me, I beseech you, Your Eminence! Lord Captain, please sheath your sword! I meant no disrespect to the Designers. Their near-zero operational failure rate is miraculous, but even 'near-zero' is not zero, especially over centuries of operation. The reliability level that they did achieve merits admiration, if not adoration."

The worldship designers may or may not use religious tracts, but they would certainly rely on *redundancy*, *diversity*, and *margin* when choosing and sizing essential systems.

Redundancy has long been recognized as a critically important design element. Indeed, the mantra of the nuclear engineer is, "Redundancy is good. Redundancy is good." Worldship designers would be expected to hold it in even higher esteem. Nuclear power plants generally have two one hundred percent capacity, physically independent groups of systems (called a "train") for each safety function. A worldship might have three or more. Redundancy allows removing a safety train for inspection, testing, and maintenance. If one train fails during an accident, another full capacity safety train is there to save the day and, on more than one occasion, it has.

Diversity is an important social imperative, but it's even more important to the concept of design. No matter how reliable a given machine may be, relying on only one design creates vulnerability to the phenomenon called "common mode failure." Extrapolating from a historical scenario, if four helicopters are needed to complete a desert operation, an inadequate sand screen design on the engine intake would doom the mission no matter if eight—or eight times eight!—identically vulnerable choppers were dispatched. Similarly, a materials defect (e.g., tainted lubrication oil) could simultaneously fail all machines that used it. Even diversity in location is important, as demonstrated during the 2011 accident at the Japanese Fukushima Daiichi Nuclear Power Station. The designers of the coastal Fukushima plant had placed all emergency power sources in basements, despite flooding being a possible common-mode failure risk. The extended duration of worldship transits would make their creators even more sensitive to design diversity.

A typical nuclear power plant taps its own generator output for normal a/c power and can also connect to the grid through a separate transformer. Worldships would similarly tap the main propulsion drive (e.g., fusion or antimatter "torch"), but by a great variety of diverse methods such as magnetic coupling, photovoltaic, thermovoltaic, and even thermophotovoltaic. The intent would be to provide multiple copies (trains) of every chosen power design, each of sufficient size to provide the necessary output. Where defective lubricant might fail all the magnetic-coupling-driven generators, the "solid state" photovoltaic trains would be unaffected. Nuclear plants supplement diesel generators with gas turbines for on-site emergency a/c power diversity. Worldship emergency power design would doubtless include multiple long-lived, battery-style fission plants, fuel cells, and the like.

Margin is another vital design element, both in building codes and operational hardware. Each nuclear plant safety train is nominally capable of supplying one hundred percent of needed power or fluid flow but, in practice, can provide more, sometimes much more. US naval history is replete with wartime stories of propellers at an rpm greater than thought possible. What those events had in common was that scared engineers called on those margins. One peacetime example reportedly took place during USS *Enterprise* (CVN-65) sea trials. Admiral Hyman G. Rickover—with one eye on the increasingly restive civilian vendor representatives on the bridge—kept adding rpm to the maximum flank bell. According to the story, after a couple of the admiral's "Two more turns, Captain," one vendor rep suddenly announced, "One more turn, Admiral, and they're your reduction gears." Rickover then reduced speed, confident that he had learned both the limiting propulsion component and its design margin.

On a worldship, the designers would craft their margins to be synergistic. For example, if the ship's radiators experienced a failure beyond their design margin, the propulsion drive would necessarily be limited to the heat output the remaining radiators could shed. Full drive thrust would be impossible. The effect would be to lengthen the trip from, say, five hundred years to seven hundred. As long as life support and other key systems had that much margin, the worldship could still safely reach its destination, albeit later than planned.

❖   ❖   ❖

Once you have calmed down, you will have questions. Hopefully, you have spared me so that I might answer them. Otherwise, you will have to summon another.

"Are you sure? How did you learn of this?"

"By the will and word of the Designers, Your Eminence." If your rank is military, I would cite the applicable standing orders. No matter what, however, my answer would be steeped in the design elements of *monitorable* and *testable*.

Monitorable systems allow operators to discern system status. Well-designed systems provide continuous affirmation of operability, and clearly announce failures or other variances from expected performance. System sensors would monitor a great many parameters. The classics of temperature, pressure, flow, level, voltage, current, etc. would be joined by ones such as continuity, tension, torque, thickness, flux, field strength, and a vast host of others. Oversight routines would interpret and weave the streams together into qualitative depictions (e.g., green, amber, and red), yet allow human inspection of the quantitative data upon demand.

Some control panels feature a layout that imitates the displayed system ("mimic bus") to simplify operator recognition. For example, plastic shapes of pipes, pumps, and turbines might depict a system, with the switches to operate valves inserted in their proper places, and with indicator lights showing position and gauges showing flow. As is the case at nuclear plants, our worldship will doubtless have systems too complicated for classic mimic bus treatment. However, the designers would know that the multigenerational nature of the ship demands a user-friendly interface. They would likely use expandable three-dimensional holograms, easily accessible and possibly even triggered by alarms.

Testable systems enable operators to determine status, whether to follow up alarms, or to routinely confirm operability in the absence of alarms. "Trust but verify" may have been a 1980s signature phrase of President Ronald Reagan, but it has always been a crucial design element. Operators are taught to trust their indications, but to verify them to the maximum extent possible. Well-designed systems facilitate both troubleshooting and operability confirmation.

Eventually I convince you that the failure is real, which leads inevitably to your next question.

"What do you propose to do?"

"It will be my duty and honor to lead the repair effort myself, Your Eminence."

While they were not successful in this case, worldship designers would strive to minimize the need for human-effected repairs. They would do so by automating as many maintenance and repair activities as possible, and by preferentially selecting *passive* (vs. active) and *static* (no moving parts) design elements.

Passive components are those that do not have to change to fulfill their design mission, while active ones must. For example, the pump that must turn on, rotate its internals, and not overheat is far more likely to fail than the pipe that will transport the pump's output. One illustration of the probabilistic difference is that US nuclear regulations require designs to preserve safety during an accident even if any one active component anywhere in the facility fails during the first few hours. In contrast, those same regulations presume all passive components remain operable during that same period. Instead, a design must be able to survive a single passive failure during the long-term cooling phases that follow, which could be months or even years.

Static aspects greatly reduce failure risk. For example, a battery that needs only a single breaker to close is far more reliable than a diesel generator that requires a great many internal moving parts to operate, as well as all the external components in its fuel and cooling systems. Worldship designers would probably make extensive use of electromagnetics and magnetohydrodynamics. Electromagnetic pumps, for example, do not rotate vanes or impellers in the flow path, but use electric power to produce magnetic forces to move electrically conducting fluids (including liquid metals and plasma). Similarly, radiators would use heat pipes, whose absence of moving parts makes them superior to systems using pumps and condensers.

Once you agree that human-performed repairs are necessary, you have additional questions.

"How extensive will the effort be and how long will it take? Is it dangerous? Are we in danger until the repairs are complete?"

My answers will vary according to the situation, of course. Except in catastrophic cases (like large object collisions), however,

I will be able to tell you that a backup system (*redundancy*) is already doing the failed system's job as I have personally confirmed (*monitorable*). Thus, the present unavailability of the broken system would constitute not so much risk as a reduction in *margin*.

"There is little danger, Your Eminence, and Scripture is clear on how to proceed."

Whether it be Scripture, Starfleet Technical Specifications, or something else entirely, a comprehensive database would exist containing repair instructions for every failure the designers could envision, no matter how unlikely. The instructions would not be limited to the spoken or printed word—languages change over time—but be in the form of YouTube-style hologram sequences. Raw materials would be retrievable, probably from vaults layered in the bow for shielding. Also in the front would be ice, not only for shielding, but also for biosphere backup and even emergency heat-sink purposes. Other items there would include spare parts, especially ones impossible or very time consuming to replicate. Fabrication facilities, such as 3-D printers and forges, would be used to produce everything else when needed.

While my answer as to how long repairs should take would depend on the specifics of the failure, they would be influenced by the design attributes of *accessibility*, *modularity*, and *standardization*.

Accessibility anticipates the need for servicing and repair, by providing spatial separation between components and an absence of physical interference. This is sometimes not achieved, most often when design modifications are made after initial installation. Late during the construction of one nuclear plant, engineers identified that component accessibility had been severely compromised in one area within the reactor building. They ended up having to compile charts listing what pipes would have to be cut to access valves in other systems deeper within the crowded compartment. Such drastic measures vastly complicate and lengthen repair activities.

Modularity simplifies maintenance and repair by grouping functionally linked components into one easily replaceable unit. It requires far less system downtime to change out a multi-component module than it does to identify precisely which individual component (or components!) has failed, gain access to it, sever it from the system, and replace it without damaging other parts in the process. In system areas involving adverse thermal conditions, radiation levels, or vacuum, swapping out modules may be the only way repairs can be accomplished.

Standardization shortens repair times because it allows a parts inventory to be practical. That is, it is far quicker to use existing spares than to fabricate each part necessary for every repair. The overall fabrication process would involve identifying the necessary stock, retrieving the materials, manufacturing the parts, inspecting the finished parts against tolerances and specifications, and then performance testing the parts. If time is particularly important, it is far superior to pull a proven part from inventory, use it, and later employ the fabrication process for its inventory replacement. Only standardizing parts can make this possible or, at the least, reduce the number of items to be fabricated each time.

You are relieved to learn that the failure has added no appreciable risk to our ship, our world. Nonetheless, you want to know how soon all can be returned to the way it was, the way it should be.

"What personnel will you use? And, when can you begin?"

My answer will be the summation of many factors, including failure extent, collateral damage, fabrication (versus replacement) needs, repair complexity, and training requirements. All those aspects except for training would be relatively straightforward, in that they could be readily calculated. How I would go about choosing and preparing personnel for executing the task itself would depend on the existing *training programs*, *simulators*, and—most importantly—on *social engineering*.

Training programs of a sound and effective nature would be a worldship requirement, absent sentient robots and/or cold sleep storage of pretrained human experts. After all, even with careers lasting fifty years, a five-hundred-year transit means that those standing watch and making repairs when the ship reaches the destination star will be ten or more generations removed from the ones who received their training before departure. Adequate training can be accomplished by a variety of approaches, including apprenticeships and shadowing, as well as by schools and testing. To sustain competence over long periods, however, programs would have to include periodic verification and demonstrations of expertise, as well as formal refresher-training periods.

Simulators would not only be the key element in achieving and maintaining expertise but would also be vital in preparing for non-routine evolutions and repair activities. US nuclear operators have benefited enormously by the federal mandate after the 1979 Three Mile Island accident for nuclear plants to have site-specific simulators. Before that, operator training relied on far less realistic methods, with perhaps a few hours on a remote vendor simulator that usually did not precisely replicate the plant that they would operate. The growth in computer power now allows current operators to learn how to respond to almost any failure in the plant. More than that, however, nuclear plant simulators have been used as a powerful investigatory tool, including verifying procedure accuracy. Worldships would have far more powerful simulators, closer to the holodecks of *Star Trek* fame than the ones of today. Such machines would be capable of simulating any place aboard the ship, allowing rehearsals of repair activities as well as control-room scenarios.

Social engineering would be pervasive in its effects, including how to prepare for a complex repair evolution. It is, quite frankly, the "long pole in the tent," the "eight-hundred-pound gorilla," or any other such analogy. Has the worldship had a stable culture throughout the long transit? After all, three or five hundred years is a long time. Is the culture a technological one? Or, do "we" live in an artificial, low-tech society, established as such in an attempt to increase stability? Maybe the ship contains cultures at multiple levels in some sort of class system. These choices matter!

Ideally, a major repair effort would involve three or more large teams, so that the work—once begun—could proceed until completion without interruption. Remote monitoring would also be continuous, as just one part of quality control and assurance activities. Materials, modules, and supplies would constantly be staged in to the work area with inspectors verifying that all is proceeding as planned. These are just some of the many jobs that would require specialized training separate from that of those actually performing the on-site labor.

How deep is the pool of technically literate and competent workers? Will the repair leader be able to simply choose from lists of qualified and experienced individuals to fill the organizational slots? Or, will repairs have to wait until enough personnel get screened for aptitude, become educated, receive basic training, and only then begin to prepare for the task?

Once personnel are selected, how many will stick out the potentially rigorous training? How many will agree to do the probably uncomfortable (and possibly dangerous) work? How will their compliance be ensured? Will they be naval officers and ratings self-selected for fidelity to duty? Will they be clerics under vows of obedience? Or, might the rewards be designed to attract the top athletes of the day?

Will the repair procedures and requirements rely on rites and liturgy? Or, would the simulators have become a central part of a freewheeling, holodeck-style gaming culture? Factors such as those will dictate how long the training will take for any evolution, including a major repair activity.

I give you my answer and, to my profound relief, you accept it.

"Very well, I approve. What are you going to do now?"

"Thank you, Your Eminence! I am off to St. Tesla's to meet with the abbot. Together, we will begin to plan the pilgrimage to the Fourth Radian Magnetic Coupling."

"May the Designers watch over you."

They have, for all these many centuries.

Nanny

Les Johnson

Les Johnson is a husband, father, physicist, manager, and author of science fiction/science fact (whew!). In his "day job," he works for NASA where he serves as the Solar Sail Principal Investigator for America's first interplanetary solar-sail mission, the Near-Earth Asteroid Scout. In addition to writing science fiction, Les writes popular science. His latest, with co-author Joe Meany, is *Graphene: The Superstrong, Superthin, and Superversatile Material That Will Revolutionize the World*. Learn more about Les by visiting his website: www.lesjohnsonauthor.com

ANGELA (AGE 9)

Nanny said it wouldn't be long until we got to go outside and play—I can't wait! Ever since I can remember, we've looked out the window and dreamed of going outside—out of the house—and into the sunshine. It's so beautiful out there. Sibby and I have thought about finding a way to open the door and sneak out without Nanny knowing, but she's too scared and I won't do it by myself. The door is too big and it might get stuck. And then Nanny would find out and we'd get in trouble. I don't like it when Nanny gets mad at me.

One day Caleb and his brothers tried to open the door and Nanny caught them. With the scariest voice I've heard come from Nanny, they were told that they were bad and that terrible things would happen to them if they went outside before they were ready. Caleb went into time-out. He stood there for an hour, unable to move except for his eyes. He looked so scared. Nanny said that it was better to be scared while you were safe inside than to be scared and likely hurt or killed outside. After Nanny let Caleb out of time-out, he ran to his room and cried. I would have cried too.

It isn't that I don't like Nanny. Or Sibby, or Kat, or Caleb, or Sadik. I like all the other kids. I just don't like being in here with all that *out there*—and not being able to see and explore it. I've watched the vids about explorers ever since I can remember. We have to. Nanny shows them to us at least once a week. I don't know why Nanny tells us how important it is to explore and then keeps us from going to the one place we must explore. All we hear is, "You will, in time."

Today I stood in front of the window looking outside for so long that my feet hurt. It was after I finished my lessons. In class, we reviewed how to grow food in the dirt, how to know when the plants need water, and what to look for to keep them healthy. Actually, planting seeds in the dirt of outside sounds like so much fun! The plants here in the house are grown in water. Nanny calls it "hydroponics." It seems that back before we learned how to grow plants in water, all we knew was how to grow them in the dirt. Now we have to go back to the old ways and I don't understand why. I can't wait to put my hands in the dirt and feel it. Nanny says the dirt outside will nourish our plants just like it does the ones that aren't ours.

Sometimes this idea of "ours" and "not ours" is confusing.

While I was looking out the window, I saw one of the many local animals climb a tree. It didn't look like any of the Earth animals we studied. It had four legs, hair, and a tail but it somehow managed to look slimy. Like a salamander, but not really. And then it happened. From up in the sky, one of the birds swooped down and grabbed it. Swoosh. That was it. The slimy animal became dinner for one of the big birds that we often see flying overhead. The birds are the only animals we've seen that look like something from Earth, only bigger. A lot bigger. They look like the drawings of flying dinosaurs, making them very scary. But I'm not afraid. Nanny says we have weapons to defend ourselves and that the birds will learn not to mess with us. I trust Nanny.

After I left the window, I played for a while with Sibby and then it was time for supper. I like supper. All the kids gather in a big, big room filled with tables and chairs and we eat together. Four hundred and ninety-two. That's how many kids are here in the house. Nanny reminds us of it a lot. "It isn't enough, but it's a start." Four hundred and ninety-two seems like a lot to me. Especially since Nanny is the only one who takes care of all of us.

Her and the housebots, anyway. Whenever one of us gets into something we're not supposed to, one of the housebots will always catch us and then will come Nanny. Whatever you say in one room, even when Nanny isn't present, Nanny knows—everything. When one of the kids is sick. When one is crying. Nanny even knew when I confessed to Sibby that I thought Zach was cute. Nanny heard us and after that we learned about reproduction. About how girls and boys are different for a reason and about babies. Nanny said we were all babies once and that she and the housebots helped us grow up. When someone asked about who our parents were, the ones who "reproduced" to make us, there was no answer. We just heard what we almost always hear: "You will find out, in time." I think Nanny just doesn't want to tell us.

"Life is fragile," Nanny says, usually after someone gets hurt and then she reminds us that there used to be five hundred of us. "Eight babies died after they were born, making you all the more precious."

After supper we had clean-up time and now it's time for bed. I'm not sleepy. All I can think about is outside and how much I want to go and explore it. But it's bedtime and the lights will soon go out. Maybe tomorrow I will be able to go outside.

MANUEL (YEAR ONE; FIRST AWAKE CYCLE)

The ship left the Lunar Gateway ten months ago and the excitement has long since faded as the reality of being cut off from home, and everyone you know there, has set in. Most are taking it well; others, not so much. I've seen tears, heard some angry words after an increasingly frustrating two-way, speed-of-light-delayed signal exchange with Earth, and a lot of people just counting down the days until they can go into cryosleep. And the eternal debate about whether one dreams in cryosleep rages on. It gives people something to talk about, I guess.

Being on first cycle, I experienced the ship's departure acceleration and saw Earth become smaller and smaller until it was finally gone. For the record, it didn't look like a pale blue dot to me. Just a dot. And then it vanished among the sea of stars. "Just a dot: not a lot," as my kindergarten teacher was fond of saying. It is amazing what you remember when you have time to reflect.

My name is Manuel Delance and I'm the ship's astrogator and first officer. Being second in command of two hundred well-trained professionals, only twenty of whom are awake on any given cycle, in a mostly automated ship, is largely ceremonial. There really isn't anyone the captain and I need to "command." For this reason, I concentrate on my *real* role as astrogator. It's my job to make sure the navigation system is working, so we can know where we are, that we're on the correct trajectory, and to work with the propulsion engineers to make any needed course corrections. So far, so good. Aside from a minor tweak in the ship's trajectory at the beginning of my shift, everything appears to be working as it should. Simply put, there was nothing for me to do.

My wife, Maria, is far busier than me. She's one of the ten doctors on board cross-trained in clinical mental health counseling. Flight rules say that one medical doctor must be awake during each one-year shift in the ten-year crew-rotation cycle. Twenty crew members are awake for a year, then asleep for nine years while the next group of twenty takes their turn. And then the cycle repeats. Each of the twenty must check in with the mental health counselor for at least one hour each week. Some take more time, but none can get away with less. Flight Rules. And the rules say that seeing the counselor is as important as meeting the daily two-hour exercise requirement. No exceptions.

Most days she says it isn't much different than being a counselor at home; better, in fact. The crew was selected partly based on passing a rigorous mental health exam—something the average client back on Earth didn't have to do. But she didn't share any details with me. She is a professional, after all.

Neither Maria's parents nor mine truly understood why we decided to leave Earth, leave them, on a one-way trip to a newly discovered planet circling a star so far away that it would take years just to send a message one way. They both wanted grandchildren, which we plan to have, by the way, but the children won't be seeing their grandparents in person—just photos and videos. We plan to

have ours as part of the first generation born when we arrive plan-
etside. It is this part they didn't understand. Why would we leave
them? Leave Earth? We tried to explain, but how do you explain
something that calls to your soul when you look at the stars. We
finally gave up and just told them we loved them.

I'm avoiding naming the planet since we, the colonists, haven't
decided yet. The nomenclature used by the astronomers is too
cumbersome and impersonal. With tens of thousands of exoplanets now known, they are mainly being numbered and given letter
designations associated with the star they circle and their estimated
similarity to Earth. B.O.R.I.N.G. Since we're the ones who will live
there and no one from home can tell us what to call our new world,
we decided to name it after we arrive. If Earth doesn't like it, well,
then they don't have to like it!

The living area on the ship is pretty big, all things considered.
Maria and I have our own quarters that we'll have to clean up
when we go to cryosleep, so the next couple awake can use them
during their cycle. By the time we are next awake, nine other
couples will have, hopefully, done the same. I imagine we'll see
new dents in the walls and some new stains here and there, but
the ship is made to be durable and resilient.

We have a galley (the food is actually pretty good), an exercise
room, a medlab, and a media room. Then there are the functional
areas of the ship like the control room and Engineering. And
the cryosleep chambers. Most people avoid looking at this part of
the ship because it gives them the creeps. Even though we know
our friends and crewmates are just asleep, they look dead. And
that's disconcerting. Maria told me that's normal.

The good news is that we have only two more months until
we go into cryosleep for nine years. I cannot imagine being awake
for the full thirty-five-year trip.

We're encouraged to keep a diary and I try to keep mine
current, but since not much happens in a given day, I'm lucky
to record something once each week. I hope this isn't getting
too redundant.

ANGELA (AGE 11)

This morning we helped Nanny birth some animals called cats.
It was just like on the videos. Two months ago, we took some

tiny glass trays from the cooler and placed them in the animal womb. They weren't much to look at—just tiny dots in some sort of gel. Each day from then until today, we paraded through the birthing room and looked at the ultrasound display to see the tiny embryos develop. They went from tiny dots to miniature creatures in what seemed like no time at all. And today they were expelled from the wombs into the padded trays we'd set up to catch them.

After we cleaned off all the gunk, a few of us got to hold them. I was one of the lucky few because I think Nanny likes me. Their tiny eyes were hardly open and they just wanted to curl up in my palms. They are so cute!

Nanny says we are going to allow a few of them to reproduce naturally so us older kids can see what that's like. Of course, we had to watch the human reproduction videos (again!) while all this was going on. Yuck. Why would anyone want to go through all that when the artificial wombs make it so clean and pain-less? Nanny said we were all born just like the cats but from the artificial human womb, not the animal one. But Nanny said the artificial wombs could only be used for us. Nanny said we would have to reproduce the "natural" way. Personally, I'd rather not.

There is something I don't understand about this whole repro-duction thing. We came from the little dishes and the artificial womb, but who put us there? Though it's gross, I can understand having a mother and being born the natural way. But we weren't. We were little dots like the cats, which means that someone had to put us there. Who? Was it Nanny? I guess it could have been, but where did Nanny get the dots?

MANUEL (YEAR ELEVEN; SECOND AWAKE CYCLE)

The boredom is beginning to take its toll. Even I am getting antsy.

This is my second time to be awake and the middle of the eleventh year of our thirty-five-year journey and there is still nothing for me to do.

Don't get me wrong. Having nothing to do should be consid-ered a *good thing*. It means that the ship's antimatter propulsion system is working as it should and that we're still headed toward our new home—an Earthlike planet circling 82 G. Eridani. The new planet-finder telescopes orbiting out near Pluto said that it is roughly the same mass as the Earth and that it is located at about the same relative position within the stellar habitable zone to give it temperatures that are approximately Earthlike. Even better, spectroscopy confirmed that it has an oxygen/nitrogen atmosphere in about the same proportion as the Earth. This last bit of news was what sealed the deal with the Interstellar Exploration Board. An oxygen/nitrogen atmosphere means there is life. Life requires a certain kind of planetary stability that seems to be rare in the universe. We finally had another target for sending people. We're still debating what to name it.

So, here we are. Two hundred colonists, zipping through the interstellar medium at a significant fraction of the speed of light, on a one-way trip to a new world that we are to make our new home and a new home for humanity. The success of the colony around Alpha Centauri buoyed the whole interstellar colonization effort and it seemed that a new ship was heading off to Centauri every year. But we want something new and different. We want to go where no one is already homesteading and tame an entirely new world. It sounds exciting. And I am sure it will be—after we arrive. In the meantime, here I am, awake for an entire year, bored out of my mind.

Maria usually doesn't talk about her counseling sessions, except in the most general terms, given patient confidentiality and all that, but the last few days she's shared with me that there is a conflict brewing between two of the crew. She won't say who, but I get the sense that it is close to reaching a boiling point. So much so that she is recommending that one of them not be awakened from stasis with us on our next cycle. She doesn't want this to come across as punishment, but more of a conflict resolution approach. I assume she'll bring it up with the captain. These kinds of decisions are made by him. At least *she's* had something new to think about.

Sometimes I feel guilty about complaining when I'm bored; the very idea of being on an interstellar starship still gives me goosebumps. Stored in the cargo bay is all the equipment we will need to build a self-sustaining colony: graphene-based housing that can be erected in hours to provide safe shelter from storms as strong as hurricanes; farm equipment, seeds, and frozen livestock embryos; several flyers and ground vehicles; a digital library that contains the sum total of human knowledge at the time of our departure; state-of-the-art chemical labs to synthesize just about any medicine or fuel we can possibly need; multiple robodocs programmed to cure just about any known disease and provide first aid for even the most critical accidents; and 3-D printers capable of making replacement parts for just about everything we have on the ship, including themselves.

Most importantly, for the long-term survival of the colony, we have ten thousand human embryos in cold storage and the artificial wombs needed for their gestation. Two hundred people, one hundred "breeding pairs"—if you will excuse my bluntness, that's not what the company calls us, but it's what we all know and consider ourselves to be—simply isn't enough to maintain genetic diversity. Even if we fornicated like rabbits, we're still limited to the good, old-fashioned nine-month human gestation period, and that means there are only so many children one hundred couples can have. And not all the children made the usual way will survive. Nor will all the original crew. Accidents are inevitable and, well, being human, so is premature death. Hence the embryos and the gene-editing system that will allow the creation of even more diverse artificially created embryos after we arrive.

The gene-editing system will enable us to create offspring that are optimized for the new planet from the cells of the original colonists and the first generation born from cold storage. There is only so much you can learn from remote sensing a planet that is tens of light-years distant. Once we arrive and assess the local environment, we can program the next generation of children to thrive in the new environment and not be hampered by our optimized adaptations to Earth—off in a distant corner of the universe.

By the way, I mentioned stasis. They put us to sleep and awaken us periodically to take turns being the crew of the ship for two reasons. The first is that no matter how good the automated system, there is no substitute for humans in the loop should something go wrong. Having people awake and available at all times is a vital part of mitigating the risk for the voyage. The second is that we don't want to sleep our lives away. You see, unlike the sci-fi vids, cryosleep doesn't stop us from aging. It just keeps us alive for the journey in such a way that we don't consume too many resources like food and air. We will age. I was twenty-seven when we left, and I will think I am thirty-one years old when we arrive because that will be the number of years I will have been "awake" since birth.

Biologically, I will be over sixty years old. Thank goodness for recent advances in life extension.

ANGELA (AGE 13)

Yesterday was Birthday One and we had a big party. Nanny said that the day the first group of us were born was the happiest in memory. Thirteen Earth-years ago, the first fifty of us were removed from the artificial wombs and put in Nanny's care. Fifty. I cannot even begin to imagine what it was like shepherding fifty babies, and then toddlers, around the house. But then I remembered that eight of the first group died, leaving just forty-two. Nanny doesn't like to talk about that and has never told us exactly what happened to them.

Just six months after Birthday One was Birthday Two, and then Birthday Three, and so on. Fortunately, those of us born first could help with the youngest ones, otherwise poor Nanny would have been completely overwhelmed. The housebots helped a lot, but, thinking back on it, Nanny did always look tired.

Sibby and I left the party early with Caleb and Thomas so we could talk about what happens next. Don't get me wrong, I think Caleb is cute, and we've snuck off before to be alone together and, well, you know, but that's not why we left early today. Even though we knew Nanny would be listening, we didn't want all the other kids to hear us talking. At the beginning of the birthday party, Nanny told all of us that shared Birthday One to be ready for a meeting tonight. A very important meeting. That's all. When we heard this, the four of us looked at each other and knew we had to get away and talk. We think we know what's coming. I'm both excited and scared.

We gathered in the hydroponics lab near the orange tree with the ripest fruit so we could snack while we talked. I am always hungry these days and oranges are simply the best snack in the house. At least to me. Sibby likes kiwi fruit and Thomas can't seem to eat enough nuts.

Caleb, always eager to take charge and lead whatever discussion he happens to be in, which is one of the qualities that attracts me to him, grabbed an orange with one hand and began talking just after we sat down near the tree.

"I think we're going to be told we need to pair off and start reproducing," he said with a deadpan and very serious face. Like he was discussing the outcome of the chess tournament we had last month. I am NOT a good chess player.

"You wish," retorted Sibby.

"Seriously, I think that's what it has to be. We are all able to, you know, and Nanny often talks about the need to get our population up." Caleb tried to look serious as he made his point, and he might have actually believed what he was saying, but I was skeptical.

"Caleb, that's all you've talked about for weeks and I think your fantasy is starting to cloud your judgment," said Sibby, grinning. It was hard to tell if she was teasing Caleb or serious. She knew Caleb and I had a thing for each other, and she might be a bit jealous.

"If you guys can get sex off your minds for a few minutes, then I will tell you what I think this is all about," said Thomas. Thomas was our thinker. He was generally kind of quiet, never the one to take charge like Caleb, but not one that was easily pushed around either. He was the kind of guy you wanted on your team during the survival tournaments Nanny had us play each week. He always had good ideas of how to survive what-ever surprise Nanny had prepared for us in the game. He was smart—and a survivor.

"The floor is yours," I said, trying to deflect the conversation away from embarrassing stuff like me and Caleb.

"I think Nanny is going to tell us we can go outside," Thomas said with authority. He looked at each of us with anticipation in his eyes, eager to see how we might react to his thesis.

"Outside? Do you really think so? It crossed my mind as a possibility, but, well, Nanny has been so adamant in saying we're not ready for so long, it just seemed impossible," I said. I really wanted to go outside and explore, finally, but, at the same time, I found the thought to be terrifying. Being outside under an open sky, in real sunshine, walking among the plants and avoiding the huge birds flying overhead is...dangerous.

"I agree," said Sibby. "That has to be it. According to the library and all we've learned about human physiology, we're almost fully mature. Back on Earth, there were entire societies that considered age twelve to be the age one became an adult."

"And the age where people got married and began raising their own children," added Caleb.

Sibby looked at him with a smirk and said, "You just don't give up, do you?"

Caleb, his face betraying perhaps a little embarrassment, looked down and began peeling his orange.

"Actually, Caleb might be partially correct. If we are allowed to go outside, it will be difficult for Nanny to keep up with all of us and, well, you know, things might happen. But that's not the primary reason for the meeting. The reason must be that we're ready to go out and do what we've been training to do. Create a village, start farming, and establish a colony." Thomas spoke with authority, like he had inside information or something.

"Do you think we'll find the original colonists?" asked Sibby, voicing the question we'd all been wondering about for years. It was a question that Nanny steadfastly refused to discuss.

"I don't know. Maybe. According to the records, there were two hundred original colonists and they must be somewhere. Why not out there?" Thomas didn't sound as sure of himself as he had previously. What happened to the original colonists had been the stuff of speculation and scary nighttime stories told by the older of us to the younger for as long as I could remember.

"Angela."

It was Nanny, calling me on the ship's speaker.

"You and the others need to rejoin the party to help clean up," Nanny said.

"We'll be right there," I replied, looking for agreement from the others. They nodded.

"Good. And don't worry; your questions, all of them, will be answered in tonight's meeting. I promise."

This caused us to do a doubletake, exchanging glances nearly simultaneously. *All our questions would be answered? All? Finally?*

MANUEL (YEAR THIRTY-ONE; FOURTH AWAKE CYCLE)

Today began just like every other day—not much happened. Sure, I checked the star charts, performed the weekly systems check, and, as the nominal second in command, I decided to run one of the Doomsday scenarios, ostensibly to keep everyone's emergency training up to date. My real reason was to give them, and me, something to do that was out of the ordinary. A way to pass what was otherwise yet another boring day. It didn't work. Sure, everyone went to their stations and pretended to be engaged with the scenario the computer cooked up—one in which we hit a small asteroid, causing major damage to the life-support systems and threatening to shut down the power system. But they knew it was a drill and it wasn't sufficiently different from one we'd run on the last shift. They did everything perfectly and averted a "catastrophic containment failure" of the power system. Translation: The crew avoided having the ship blow up with their customary expertise and efficiency. This was, after all, a ship crewed with the best and brightest from throughout the solar system.

Midday, we had some excitement. We received a new data dump from Earth. At these distances, two-way communication of any sort was simply impossible. Our distance from home was now measured in light-years—meaning that any message we or they send will take years to reach its destination, making a conversation impossible. Instead, our ship's computer sends weekly status reports back to Earth using the ship's optical communications laser. By the time the laser light reaches Sol, the beam has diverged considerably but not so much that the kilometer-scale receiver can't get enough of the signal to understand it. Communication across large distances was like that. On each side, you needed either a large antenna array or a lot of power. Earth had both, so they could broadcast using extremely powerful solar-pumped lasers or receive using antennas that were hundreds of kilometers across. We had power, thanks to the antimatter system that drove the ship, but our receiver was limited in size. It worked, but with ever-worsening performance as our distance from home increased.

From Earth, the communication is not so frequent. At first, we carried on conversations despite the minute-scale and then hour-scale time lag. When the distance caused the time lag to become days and weeks, the update frequency slowed accordingly. Now that the time lag is measured in years, we are lucky to get an update more than once per month.

Today's update brought news that the Alpha Centauri colony had reached a milestone—one hundred thousand people now lived there. This was enough for them to declare that humanity had a new second home and was finally out of the cradle—and eliminated the risk associated with being a one-planet species.

We also learned that another possible Earthlike planet had been discovered around a star only about five more light-years from Earth than our destination. No one had announced plans to go there, but it was only a matter of time. We also learned that at least two more groups announced plans to follow us. They might launch within the decade, meaning that they might arrive in my lifetime—when I am in my eighties. Even with life extension technology, that was up there. I wonder what they'll find when they arrive. Will I still be alive? My children? Or will we all be dead, killed by some alien plague? Best not to dwell on the negative.

By the end of the day, it became clear that this was not just another boring, uneventful day. And that was unfortunate.

Deputy Chief Engineer Jeremiah outright slugged Captain Tsuda in the mess hall at supper. We all knew Jeremiah and Tsuda didn't like each other. That became apparent on a previous shift during their frequent shouting matches that had resulted in Jeremiah being reprimanded and, supposedly, swapped in cryosleep for Deputy Chief Engineer Marsee for the remaining awake shifts. That happened last shift, but, as is I supposed inevitable, the wakeup schedule wasn't changed for this shift and both Jeremiah and Tsuda were again awakened at the same time to serve together.

Jeremiah didn't just slug Tsuda, he knocked out his two front teeth and gave him a concussion. After that, Jeremiah ran to the Propulsion Engineering room and barricaded himself in by shorting out the control circuit on the access door. We had a crazed engineer on the loose. Aboard a starship, a crazy engineer is a scary thing indeed.

The alert went off, just like it did in my simulation earlier in the day, but everyone knew that this time it was not a drill. Amid the scurrying of crew to their duty stations, I managed to speak briefly with my wife, Maria, just as she arrived in the medlab. Maria and I spent most of our time awake with each other and I knew she was, until now, just as bored with her days as me.

"Manny, they are bringing in the captain. I'll let you know how he is as soon as I check him out. Don't worry about me. Someone needs to keep Jeremiah from hurting anyone else."

Until this moment, being second in command of the crew during a shift was more ceremonial than anything else. With nothing happening, Captain Tsuda and I really didn't have much to do and neither of us, as civilians, were big on formality or protocol. But now, as Maria just reminded me, it was up to me to figure out what to do with Jeremiah and secure the ship.

"Maria, you take care of the captain and get him back on his feet as soon as possible. I'll see what I can do about Jeremiah," I said, getting my sea legs as acting captain. I surveyed the room and the expectant faces staring at me. This was definitely different from the simulation.

"Kearan, you and Francesca get the stun guns and come with me to Engineering. We need to talk Jeremiah out of there before he does something else stupid," I said as I moved toward the exit and the corridor that would take us to where Jeremiah was holed up.

When we arrived, we were met by Jeremiah's supervisor, Emory Vulpetti. Vulpetti's face said it all—we had a big problem.

"Emory, I'm glad you're here. What can you tell me about Jeremiah and what he's up to?" I asked.

"Take a look," said Vulpetti as he handed me his tablet computer. On the screen was a live video feed of Engineering that showed Jeremiah busily adjusting one of the control panels. He seemed frantic.

"Do you know what he's doing?" I asked.

"I can tell you exactly what he's doing. My tablet can monitor every workstation in Engineering, including this one. He's trying to lower the antimatter radiation shields."

"Does he want the ship to explode?"

"Apparently not. He just wants to kill all of us. The shields he's lowering are those that shield the living areas of the ship from the radiation produced during the matter/antimatter annihilation. When matter and its antiparticle collide, they cancel each other out and release energy—that's what propels and powers the ship. The energy is in the form of charged and uncharged pions, neutrinos, and gamma rays. The bottom line is that without the shielding, we'll all be fried and die from the radiation. Quickly."

"Can you stop him?"

"Not unless I get in there. All I can do is watch unless I'm at the controls. He's locked out my remote access to this workstation."

"Well, dammit, let's find a way to get you in there. Kearan, you and Francesca help Emory get that door open while I try to talk sense into Jeremiah."

I reached down and activated my comm, hoping that either Jeremiah would listen to reason or that I could distract him long enough for us to figure out how to get in there and stop him. When I spoke, Jeremiah looked up at the room's camera.

"Jeremiah. This is Manuel. It looks like you're having a bad day. Let's talk about this and not make it any worse, okay?"

"Manuel, just go away. I've had enough of you, of Captain Tsuda and all the rest. What the hell are we doing out here anyway? We're bringing all our bullshit from Earth and spreading it to the stars, that's what. We're not worth it. We're not worth saving. One of these days the people on Earth will finally destroy each other and then it'll be over. Hopefully someone on Centauri will see the light like me and stop the cancer from spreading there too. I'm going to do my part. You, me—we're done. You just don't know it yet."

Jeremiah was truly over the edge and at this moment I desperately wanted Maria to be here. She's the trained counselor, not me. If anyone can talk Jeremiah off the ledge, it will be her. I cut off my comm with Jeremiah and paged Maria.

"Maria, get over here now. Jeremiah is about to flood the ship with radiation and you're the only one with the training to keep him from killing us all."

"I've got the captain stabilized. Natalie can take over and not miss a beat. I'm on my way," she replied.

"Emory, is there any place on the ship that would be safe from the radiation if the shields go down?" I asked.

"Only one. The cryostorage chamber where we keep all the embryos. They are extremely fragile and highly susceptible to radiation damage, so the chamber has a ton, literally, of extra shielding," he said.

In my mind I visualized the cryostorage chamber and the freezers there. It wasn't very big. Big enough for perhaps five or six people, but no more. And there would be no room for supplies. Not good.

"Will any of the spacesuits provide shielding from the radiation? They're good enough for going outside to make repairs and be exposed to cosmic rays."

"No good whatsoever. The flux of annihilation byproducts is huge. Against that, the spacesuits are useless."

"What about the cryosleep chamber? Can we put everyone to sleep in there until we get this mess sorted out?"

"There's not enough shielding. All the sleepers will be fried just like us."

I looked up just as Maria rounded the corner. Thank God. Maybe she could talk Jeremiah out of killing us all.

"I've been listening. You left your comm open," she said, looking at the screen as Jeremiah continued to tweak various controls on the panel in front of him.

"Jeremiah, this is Doctor Delance. I'd like to speak with you," she said in her calm and professional voice as she looked anxiously at the screen.

"I figured they would call you. Maria, I like you. But you're ultimately no better than the rest of us. Selfish and evil at your core. We're parasites on the galaxy. We almost destroyed the Earth and we will do the same no matter where we go. I'm sorry, but you'll have to die like everyone else," Jeremiah said, looking up at the screen and taking a brief pause from his suicidal efforts.

While Maria engaged Jeremiah, Emory motioned me to one side.

"Manuel, you need to get to the cryostorage chamber with my tablet. From it you can monitor and control just about every system on the ship. You can make sure we stay on course for 82 G. Eridani. That's your expertise. You can make sure the ship gets there even if we can't. Just in case," Emory said as forced his tablet into my hands.

"I thought you said he'd locked you out?" I asked.

"And why can only one person do a lock out?"

"Safety measure. We'll fix it later. Right now, he only locked me out of the remote control for the antimatter flow and shielding. With this we can still access propulsion and navigation. From what I can tell, we've only got another two minutes before the shields go down. You'll have to go now."

"But I can't. You need to go, not me. You're the chief engineer. You can keep the ship running better than me."

"No, I can't. The ship is mostly automated. The crew is supposed to be the *redundant* system. The 'safe' backup in case the automated systems fail. The ship is running fine. The only thing that might need tweaking is the trajectory. You know that. With the distances involved, even a small calculation error on one end can end up being billions of kilometers worth of error on the other and we can't afford that. That's your expertise, not mine. I know you're in charge with the captain out of commission but quit arguing with me and go now."

I looked at Maria. I looked at her because she was the most important thing in the universe to me at that moment and I didn't know what to do.

She looked back, her brown eyes becoming glassy, and said, "I'm not having any luck with Jeremiah. You need to do as the chief says and go. We've got to save the embryos and the mission. Manny, go now."

I grabbed the tablet and left. That was the last time I saw my beloved Maria and any of the crew alive.

For two days I was alone, "safe" among the frozen embryos in the most heavily shielded part of the ship. Everyone else died. Most within minutes of the shielding going down; others took a little longer. I forced myself to watch, knowing that what I did in the time I had left could make the difference between life or death for you, the ship's most precious cargo. Though there was no food, there was enough water and air in the cryostorage chamber to sustain me a few days—long enough for the system to perform automatic reset and restore the radiation shielding.

I took the bodies, placed them in the airlock, and spaced them. That's what I would have wanted, and I had to assume the same for everyone else.

I checked and rechecked all the ship's automated systems, our trajectory, and the systems needed to land and build the first habitat upon arrival at 82 G. Eridani. Unless something else catastrophic happens, you should be safe and become the first generation of humans born there.

The nanobots—nannies, as the crew liked to call them—are programmed to take care of all your physical needs after you are removed from the incubators. They can feed you, change your diapers, and patch you up as you learn to toddle and have the inevitable childhood injuries. But they can't nurture you. They can't hold you, look in your eyes, or teach you how to talk. They can't *nurture* you. We hoped to parent you, to be your real nannies, as you grew to be the first generation in our new home.

I'm the only adult left that can guide you. Can I do what needs to be done?

Can Manny be your daddy and your nanny? I mused as I found a way to chuckle at the wordplay and smiled, despite the circumstances.

The way I see it, there are three options. Option one: Put myself in cryosleep, trust the automated systems, and wake up upon arrival. Option two: Stay awake for the four years remaining. Option three really isn't an option. I refuse to let Jeremiah win by killing myself or putting you at risk. I'm afraid to put myself in cryosleep; what if something happens when I am asleep? But I'm also afraid I'll go crazy if I try to remain awake, alone, for the rest of the trip.

Whatever happens, please know that life is good. You are good. And don't let the Jeremiahs of the world tell you otherwise.

ANGELA (AGE 25)

Our colony on New Hope—that's what we decided to call our world—was thriving. All the original embryos were now adults and reproducing the old-fashioned way. Nanny, no, Manny—I really should call him by his given name—was with us until last year. He helped us at every step, but his age really started catching up with him. I was with him when he died, and it was just before then that he shared his diary with us and I learned about his beloved wife. I was thirteen when we learned his real name and some of the story about the disaster that overcame the original crew during their journey here. He never told us that the crew was murdered. He never once mentioned his wife. All he would say was there was a radiation leak and everyone else died. I guess the memory was just too painful.

I cannot imagine what it must have been like to be on the ship, alone, for four years. He must have been so lonely. I think we are what kept him sane. He willed himself to live for us, though we were nothing more than tiny embryonic dots in freezers. We all cried.

We now have a town, our first elected government, and a thriving farm-based economy. The 3-D printers are turning out just about anything we need to have and we're mostly healthy. The pterodactyls, as we're calling the native birds, are pretty smart. Once they learned that we could, and would, fight back, they mostly steer clear of us and our growing community. We haven't seen any signs of emergent intelligent life—at least, not yet.

Caleb and I decided to name our first child Maria.

She was born just before we received word that the next colony ship was to arrive within the year. All of us were excited to learn that people from Earth, people born on Earth, would soon be here among us. We have so many questions that Manny wouldn't answer and that can't be answered by looking at library texts and films. So many questions. I hope they don't mind.

Those Left Behind

Robert E. Hampson

Robert E. Hampson, PhD, turns science fiction into science. With over thirty-five years' experience in neuroscience from animals to humans, he recently led a team which demonstrated the first neural prosthetic to restore human memory using the brain's own neural codes. He has advised more than a dozen SF/F writers, as well as game developers and TV writers via the National Academy of Science's Science and Entertainment Exchange. Dr. Hampson writes both fiction and nonfiction for SF/F audiences, ranging from military SF to Zombie Apocalypse. Some of his work previously appeared under the pseudonym "Tedd Roberts."

Melisande "Mace" Wolfe blinked and checked her time display. Sixteen thirty-two. The shuttle to Midland Spaceport was at eighteen hundred, and there was an hourly Tube to F'burg. She adjusted her chrono to groundside time. After six hours of transit, it would be near midnight Universal Solar, but 6:00 p.m. Central time when she arrived at her parent's house in Fredericksburg, Texas.

It was only eight years since she'd left, but she was amused at how Groundsiders still set their time by position of the sun in the sky. She'd spent most of those years using the sun as a fixed navigational point, not a timekeeper.

She lifted her arm and read the message on the cuff display: "Agreed, then. One last Thanksgiving with the parents. Bring earplugs and plate armor. —Sandy"

201

Earplugs and plate armor. It was hard to tell if that was a warning or humor, but it fit Mace's last memory of her father screaming and throwing a half-full beer bottle in her direction as she left the house for the last time. Mace—never Milly or Missy—and her big brother Alexander, whom Mace affectionately called "Sandy," graduated college at the same time and they threw a celebratory party that ended badly. Mace left that night, abandoning most of her belongings in the process; she heard that Sandy had packed and left the next morning.

Although three years younger, Mace caught up with Sandy in college by taking—and honoring in—an extra load of courses most semesters in both high school and college. It hadn't helped at home, though; nothing Mace or Sandy did ever seemed to please their father. She supposed there was a time when Sidney Wolfe loved his family, but that time was long gone. Holidays were a particular strain, given that half the time Mom wasn't even there. Melanie Wolfe had been an on-again, off-again mother, spending weeks to months at a time pursuing "her arts"—which usually meant performance gigs, bed hopping, artist communes, "trial separations," and periodic stints in drug rehab. The four of them hadn't spent any time together since the graduation party.

Mace lifted her go-bag and secured the snapstrap to the socket on her left shoulder. The weight was negligible; Mace still didn't have much in the way of belongings. That was not due to change at any time in the future, either. She noted that there was still time for dinner at the station before the shuttle. Boarding wouldn't happen prior to fifteen minutes until departure, and long lines at security were a groundside affectation. Mace would flash her PCbCorp ID and walk on board. Yeah, best to eat now—Sandy promised to cook tomorrow, but the fridge wasn't likely to contain much except beer tonight when he arrived.

Mace could see the shuttle unloading across from her seat at the station restaurant. She supposed it was a rather cruel joke to the newcomers; they were unlikely to want food after the docking maneuvers *or* their first few moments in the rotating habitat. In fact, she could see some newbies among the people returning to Gilster Station. The veterans had a stiff-necked manner, slowly rotating their whole bodies instead of turning their heads; well-informed newbies bought collars that provided only limited rotation at the neck, and she could see a few of those. *That guy,*

though ... no collar, wide eyes, he's going to look around, I'll bet!
Sure enough, the passenger turned his head to look around. Almost
as soon as he did so, he fell and started puking. Rotational gravity
was like that. The human vestibular system could adapt to any
motion, but novices didn't take into account that in the constant
angular momentum used to simulate gravity in a rotating space
station, turning your head quickly meant your brain thought
you were suddenly accelerating sideways. You learned to avoid
sudden turns, wore one of the restrictor collars, or you got an
interrupter implant like Mace. There were other treatments, too,
but they required extensive physiological modifications.

The gate agent was there right away with a collar and medpatch
to help the passenger. A couple of hand-sized bots were already
cleaning up the mess. There was at least one newbie on every
flight. By the time she boarded in—she squinted and checked the
time display—in five minutes—there would be no evidence of the
accident. Mace rotated her wrist and looked again at the message
from Sandy. She could have just transferred it to her heads-up
display, but she preferred the screen built into her left wrist.
Social custom, she supposed. Most people got their messages on
a wristcomp, so Mace configured her forearm accordingly. Social
custom was still important to her, although she supposed that
was all about to change.

Alexander Wolfe paid the autocab and pulled his Plac and the
grocery bags out of the storage box on the back. The Personal
Luggage Allotment Case had straps to fit over his shoulders, and
he distributed the mesh-and-plastic bags between his hands. The
house in Fredericksburg hadn't changed much in eight years.
Flaking paint, broken shingles, chipped mortar and bricks all just
as he remembered. He could still see the loose security bars on
the single window over the garage where he used to sneak out at
night. The rose bushes he planted for Mom were gone, but then
they'd been dead by the time he left.

Sandy thought it unlikely that the sensors on the doorknob
would recognize his ID. It was eight years, after all, and he'd had a
few ... modifications ... since then. For that matter, he'd needed to
reset his biometric signature every few weeks for the last couple of
years. No, he would ring the doorbell and hope it worked—or knock
and hope someone was awake enough or sober enough to respond.

The bell worked, but no one answered. He could check the lock, but he didn't have a hand free.

He looked around. The neighborhood was run-down. Tall weeds and broken windows suggested that half of the houses were empty. In fact, his parent's house looked positively well-kept in comparison. Good. There wouldn't be anyone to react badly to what he was about to do.

He looked down at his feet. He preferred to go barefoot most of the time because conventional shoes were too uncomfortable. His skin was tough and resistant to almost anything that he could encounter in space; rough ground, sidewalks or even the gravel streets of his old neighborhood didn't bother him. He needed some way to keep his feet clean though, so he wore slippers when necessary. They looked like thin black gloves rather than shoes, but they were more comfortable that way. Sandy balanced on one foot and reached the other up to first slip off the "glove" and then grip the doorknob. As expected, it didn't turn—his galvanic response and DNA were too different now. It had a keypad, though, and Sandy was certain that the combo would be the same. Mom would never think to change it, and he was sure that Dad assumed that he and Mace would never be back barring the end of the world.

Sandy worked the combination with the digits on his foot. It was no longer accurate to call them toes, although one of his coworkers preferred the term "tingers." The code worked and the door clicked open.

"Mom? Dad?"

He pushed through the door and worked his load through the front hallway, across the living room, and into the kitchen. He sat the bags on the counter and started to unload the contents. As he looked around, he noticed that the kitchen—in fact, anything he could see of the house—was surprisingly neat and clean. *Hmph. They must have a maid come in. Either that or I'm in the wrong house!* The kitchen had the same old look, though: a wall clock he made in Scouts, and the ladle holder Mace made in elementary school. It was the right place.

The fridge was relatively new, and despite his expectations, had food in it—not just beer. Most of it appeared to be quik-heat and zap-wave meals, but it was food.

...and no beer. Strange.

Sandy started putting away the food he brought. Fresh, unfrozen turkey, all the makings for stuffing and gravy, fresh-picked vegetables, fresh fruit and dough for pies. He would cook tomorrow—well, in fact, he should get a few things started now. He *liked* cooking and his friends said he did it well. Of course, he insisted on fresh ingredients. *The curse of being a life-systems bioengineer, I suppose.* Dad didn't cook, and Mom's forays in food preparation were usually influenced by whatever was in vogue in her artist community. Most Thanksgiving meals were best forgotten, such as the year she prepared a "deconstructed raw-food feast" consisting of flavored soy chunks for the turkey, raw corn (still on the cob!), wheat germ in place of stuffing, and pumpkin seeds for dessert. That wouldn't be the case for *this* meal.

He pulled some pots and pans out of the cabinet and wiped the dust from them, drew some water, and sat them on the stove. He tried the turn on the stove, by pressing the igniter, but it wouldn't take. *I guess Mom never had a reason to replace this antique.* Natural gas was cheap, too, and Dad liked cheap.

He reached into his backpack. There were surprisingly few clothes and quite a few cooking utensils. He decided before leaving the station that he would need to bring his own. He pulled out a small utility melder—useful for plumbing repairs and spot-heating. It also made a fine igniter for almost any fuel source, but even that didn't ignite the stove. He could smell the gas, but it wasn't much, so it was probably just a bad supply. It was obvious from the appearance of the neighborhood that maintenance was a low priority here.

He turned off the gas and reached into his pack again. He pulled out a rolled-up mat and snapped it to create a rigid, half-meter square that he placed over the stove. With a touch on the edge, he activated two of the four radiant heater spots on the portable cooktop. Oh yes, he would need those supplies to prepare a classic American Thanksgiving Dinner such as he and Mace always wanted, but never had.

Mace decided to walk from the tube station. It was 5:30 p.m. local time, and just getting dark in south Texas. The air was cool and dry, one of the virtues of autumn in the south, and this was her last time in the outside for . . . oh, about the next hundred years. Of course, she would hibernate most of that, unlike Sandy who would be awake for twenty years during the trip.

It came as a great surprise to them both when the roster for *Centauri Dreams* was announced. Mace was unaware that her brother applied to the crew. She tried to keep in touch with her big brother for the first few years after college, but flight training and missions to the Belt made regular communications increasingly difficult. When the list was published last April, she was surprised to see her brother's name above her own. It surprised Sandy, too, considering the email she received about thirty minutes after the list was out.

"Hey, sis, long time! I'm glad to see you made it. It's a small universe. They're sending me to the *Dreams* for final checkout next month. When you pass through Gilster, give me a call." He included a comm address and priority contact code. It was a Proxima Centauri b Corporation code. Mace wasn't due to transfer jobs to PCbCorp for at least another four months, so it was doubtful that she would be at Gilster Station until right before departure next year. Those final months would be hectic, and she doubted that she would have enough free time to spend with her brother when the time came. It didn't mean she couldn't write, though, and they were in the habit of corresponding every week. The irony was that she arrived at Gilster a month ago, but Sandy was already groundside for the last of his "adjustment" sessions.

Mace turned onto the street she once called home. It was more run-down than she remembered. Fredericksburg was never a large town, just support for the outlying ranches and a historic downtown square. It became popular when the electronics industry started getting new deep-space contracts. The San Antonio–Austin corridor was overcrowded, and residents enjoyed the respite that being outside the urban corridor had to offer. A small aircraft company in nearby Kerrville picked up a contract for aerodynamic control surfaces on shuttles operating out of Midland, and the San Antonio–Midland hyperloop passed right between Fredericksburg and Kerrville, just five miles from either city. The influx of jobs combined with people who could telecommute to high-tech jobs while living on the edge of the Texas Hill Country. The entire region started growing when she and Sandy were young; but this particular street was none too popular given that it was just the other side of a dry creek bed from the wastewater treatment plant.

The old house still looked the same, including the broken fence rail where she repeatedly tried to set off a model rocket made with old fireworks. She was in middle school and interested in space, inspired by the many neighbors and classmates whose parents worked in the supporting industries. Fortunately, she was still interested in space even after getting yelled at by her father.

The front door was unlocked, and she smelled food cooking. "Mom?" she called. She supposed it could be Mom, even though growing up her mother's idea of a cooked meal involved someone else doing the cooking.

"Nope, just me, sis," called Sandy's voice. "I don't know where they are, but I figured I'd better get started. Still only one oven and the stove's just as crappy as the last time."

"Yeah, I remember that. Not that it ever got used much," Mace answered. "So, where's the rents?"

"Don't know. Dad's side of the garage is empty, but there's fresh oil, so that's a car gone. Mom obviously still has that roadster; it's on the other side on chocks with the wheels off. No idea where they are. The house was locked, but the combo was still good."

"Strange. I suppose Dad could be at the VFW bar." Mace paused a moment. "They *do* know we were coming, right, Big Bro?"

Sandy turned, and looked right at his younger sister just before he swept her up in a bear hug. "Yeah, they know. I wrote; got an acknowledgement. I even commed Mom a few times over the last couple of months. She said they were looking forward to our visit and even had a big surprise." He disengaged, stepped back and shook his head. "She sounded strange, though, more serious. I'm not sure what that's about, but she sounded good."

Mace grinned. She and Sandy were always close. Most of the time, they were each other's only support, especially when Dad was raging and Mom was absent. Funny, though—she remembered Sandy being taller. *Have I grown that much? Is it one of those side effects of free fall that they warn us about?* She stepped back and looked frankly at her brother and compared his height to the cabinets and fixtures in the kitchen. *Yeah, definitely shorter.*

"Okay, bro, my eyes are working okay, and I can tell that you're about five to six inches shorter than you used to be. What gives; I thought free fall was supposed to make you taller? You've been doing a bunch of zero-gee stuff, right?"

Sandy grinned. "You should talk, sis. I felt plastic and titanium with that hug." He sighed. "No, it's the mods. The docs actually wanted me to be even shorter so that the free-fall stretching didn't affect the heart too much, but this is all from the hip and leg mods." He balanced on one foot, lifted the other, and waved it around just like a hand.

As he put it back down, Mace could see that Sandy stood a bit bowlegged. His hips were wider, but with less bulk. His thighs were thinner, in addition to the broader feet and longer toes. "Ah, I see. So, you've got four points of contact in zero gee—well, five, when you land on your butt," she teased. "Doesn't it hurt in full grav, though?"

"Not really, they reengineered the pelvis and femur. Shifted the head of the 'socket' for more flexibility, but also added a bone extension that locks into the pelvis when I need to stand. My shoulders do it too, which lets me brace in full or partial gee with better leverage. The new toes are great, too!" Sandy lifted his foot and slapped Mace's hand where she was digging into the grocery bags.

"Ouch!" Mace said with a hurt expression, then grinned. Sandy slapped her left hand. "Actually, though..."

"Yeah, I've got full sensory in my tingers, sis. Titanium, polymer... glass?" He looked at Mace quizzically. "How much?"

"Silicon nanotube. Standard pilot hemicorporotomy: left side from shoulder to feet, plus left eye and ear. Cyber-neural mesh on the right hemisphere." Mace shrugged as if it was no big deal. After all, her genes were still the same and she still had her—admittedly short—hair, unlike Sandy.

"The eye I can see. Gramps would have loved that electric blue. The rest isn't obvious."

"Yeah, well there's a covering I can remove to allow electro-optic interfacing if the inductive and IR pickups don't work." Mace dug her finger into the left wrist to peel back a portion of the skin-like covering.

Sandy saw metal, ceramic, wire and optical components reflecting in the overhead lights. His own specialized vision allowed him to see deeper into the prosthetic, revealing multicolored optical nodes ranging from infrared to ultraviolet. "Cool. What if you're left handed?"

Mace set the "skin" back in place and it closed without a visible seam. "Then they do the right side and left brain." At Sandy's questioning look, she continued. "It's not a left brain-right brain thing, that's been largely disproven. It's more that the hemisphere opposite to your handedness tends to be less specialized. Left brain controls the right side, right controls the left. The less specialized side is plastic enough to support additional functions. So, if you're right-handed, you get left-side bionics and right-side cyber mesh to support it."

"Damn. I thought what they did to *me* was invasive. How bad was it?"

"Hurt like hell until I learned how to control it. There's a bed at Selene that's been crushed into a one-meter ball from when I got mad at the docs."

"Heh, yeah; been there, done that. Well, not exactly, but apparently I smashed three nurses in the face, broke some bones and loosened some teeth because I was flailing around and locked my joints." Sandy looked down in shame. "I was in pain, but that's no excuse. They'd just adjusted my femurs and it *hurt*. Felt pretty bad about it afterward."

Mace found the soft drinks that Sandy placed in the quik-chill compartment of the fridge. She opened one and took a sip. Mmm. A local pop flavor from their childhood. She missed this, and it was probably the last chance to indulge during this quiet lull before the ordeal that was to come. She held it up as if in a toast. "Here's to us, bro. There's none like us!"

"Hear, hear," Sandy replied.

"Shouldn't someone be home by now?" Mace asked after taking another sip. Almost immediately they heard a vehicle crunching down the gravel of the street, then one, two, three, four car doors. The siblings looked at each other.

It was their parents, accompanied by a couple midway in age between their own ages and that of their parents. The man and woman were introduced as "Brother Erebus" and "Sister Elizabeth." It was not immediately clear if they were husband and wife, actual brother and sister, or some other relationship. They were strangers to Mace and Sandy, and not the only strange thing going on.

The first thing that Sandy noticed unusual was that Dad was in a button-up shirt and tie. There was no smell of booze and he seemed somewhat civil. Mom was dressed neatly as well—and she hung on Dad's arm. They looked like a textbook example of respectable middle-aged husband and wife—which was totally unusual for them. They seemed genuinely glad to see their son and daughter. Sandy had the overwhelming sensation that something was amiss.

The guests were an entirely different matter. Sure, they dressed nicely, Brother Erebus ("Call me Eric") was wearing a suit. He spoke all the right words about being happy to meet "Sid and Mel Wolfe's famous kids," but there was something in his manner that just wasn't right. "Sister" Elizabeth appeared anything but happy to meet them. Her gaze lingered for a few minutes on Mace's face, likely taking in the startling blue of her artificial eye. When she looked at Sandy, however, she let slip a look of pure hatred before replacing it with an expression of disinterest.

Sandy offered everyone drinks and some light snacks he had picked up just in case they needed something this evening. Surprisingly, the requests were for soft drinks and water all around. *What? No beer for Dad? What the hell was going on here?* It didn't take long after the introductions for Dad to explain:

"Alexander, Melisande...I'm ashamed of my behavior...of how I raised you. I'm not proud of my drinking or my anger and I want to apologize to you both right now."

"Um, gee, thanks, Dad, but how did this come about?" Mace had a frankly incredulous look on her face.

"As you know, I'm a drunk and abuser. Most of the time when you were growing up, I was drunk. I yelled, I screamed. I often raised a hand, although I never struck either of you or your mother."

"You sure threw things, though," Mace interjected with a hint of bitterness.

"Yes, Melisande, and for that I am truly sorry." He looked briefly at his wife, who nodded. "That changed about two years ago when I struck your mother." He held up his hands to forestall their comments. "Oh, not with my fists." He looked down, and Sandy could see the beginning of tears in his eyes. "I did it with the car. I was angry. I'd been drinking of course, and I was careless. I was pulling out of the driveway too fast and your mother came out to try to stop me. I struck her with the car."

At Mace and Sandy's shocked looks, their father let out a sob, and their mother reached over to lay a hand on his shoulder. After a few moments, he collected himself and continued. "So, I decided to turn my life around. I spent a couple of months in rehab and started counseling—your mother, too. In the last year we started in at the Grace and Unlimited Potential Foundation down in San Antonio. It's just a few stops down the tube line into the city. We met Brother Eric there not long after we started, and he has taken over the counseling. He was so excited to hear that our family was coming for the holiday and that we would have this chance for forgiveness."

Something in the story seemed a bit...off...to Sandy, but he couldn't put a finger (or tinger) on it, so he made no comment. Still, the name of the church, or club, or whatever it was, seemed familiar. After more confession and contrition on the part of both parents, and signs of a little bit of thawing on the part of Mace, Sandy mumbled his own acceptance and that he really needed to get back to food preparation.

As if that was their cue, the couple from the "fellowship" said their goodbyes, saying that they wanted the newly reunited family to have time to catch up. They accepted an invitation to dinner the next day. Sandy could hit the local grocery in the morning for just a few items to stretch the meal to feed two more. It's not that he overbought in the first place—that was a bad habit in space—it's that he planned on packaging the excess and taking several meals back to his colleagues. He also knew efficient and nutritious ways to stretch rations; so, there was always a backup plan for getting the highest yield from his ingredients.

The evening ended quietly. Sandy went back to cooking and their parents said that they needed to perform their evening devotional study. That left Mace to either wander about alone or retreat to her old room and try to sleep. She stepped out into the backyard and looked at the tree she used to climb in her younger days. It seemed so much bigger back then. Even though the ancient ash tree clearly increased in girth over the years, it was no match to her memory of it. In her early teens, she'd built a small ladder to help reach the lowest branch, climbed up the sturdy trunk to the upper part, then laid back and looked at the stars. These days, she could just jump the entire height if she wished—balancing on her left leg and taking advantage of the superhuman strength built into the prosthetic. It was more fun to climb it, though, so she did.

The branches might have been thicker, but she still heard

some ominous creaking as she climbed. Well, she was bigger, and her body was denser, thanks to her prosthetics. The tree was old when her father grew up in this house, so that made it... wow! Well over sixty years old. Maybe she shouldn't climb all the way to the top. At least the leaves were down already. She settled for a sturdy branch about halfway up and leaned back to look at the sky. It was still the same broad expanse of Texas sky that thrilled her as a child, perhaps a bit brighter since the surrounding cities became smarter about lighting.

Sandy woke up early and walked to the grocery store about a mile from the house. His father offered the use of the family car, but Sandy demurred. It wasn't that far to walk, and he needed to enjoy the sun and fresh air while he could. Besides, he learned last night that his parents had volunteered to serve the Thanksgiving meal at the G.U.P.F. mission house down in San Antonio. Oh, he or Mace could have driven them to the Loop station, but they both agreed that this new contrition on Mom and Dad's part was just a bit unsettling. Being in the same car would be too awkward right now.

Mace was still sleeping when he left the house—she'd stayed in that tree until well after midnight. His parents would likely be gone by the time he came home from the store, so he would have plenty of time to cook and work through the conflicting thoughts that disturbed his own sleep. This new situation was just too unexpected. The person who answered to "Dad" bore little resemblance to the person he knew when he left home. There was a time when Dad was different—smart, funny, less angry—teaching science in the local high school. Mace was too young, but Sandy remembered riding on his father's shoulders as they toured the Alamo and went to museums in San Antonio, Houston, and Dallas. That was before the school district realignment that sent Sidney Wolfe into inner-city Austin and started the downward spiral of depression, alcohol, and anger that drove their mother to spend more time away from home than with her son and daughter. Sandy grew up with a dysfunctional family for many more years than a happy one, and Mace never really knew the good times.

Now, though, it seemed as if their parents were polite strangers answering to the same names, but not someone Sandy really knew. Could their new "fellowship" have brainwashed them or be controlling them somehow? No. That just wasn't possible. He and Mace were both trained to spot abnormal and artificial behaviors—it was just too dangerous to ignore those signs in a space crew—and there was nothing artificial in Mom and Dad's words last night.

Brother Eric and Sister Elizabeth, though—they were a different matter. Something about them disturbed him and touched right on the edge of those behavioral alarms in his subconscious. *The name of their "fellowship"... that seems familiar somehow.*

Mace awoke to a quiet house. 0800 wasn't all that late by local time, although if she were on Gilster, she would have missed half of her shift by now. Sitting in the tree last night helped her adjust her schedule, and sunlight was a powerful influence on circadian rhythm even though she had artificial means to adjust her body's internal schedule.

Her parents mentioned going into San Antonio for "mission work" this morning and Sandy was headed to the store for enough food to expand the menu from four to six people. He would be back soon, but for now she was alone in the house. Somehow it still felt cramped. She flew shuttles and lived in a twenty-five-cubic-meter cabin, yet this house felt small. She called up a calculation function from her cyber interface, the Telencephalic Heuristic Neural Grid-Enhanced, THNG-E, or the "thingy in the brainy" as Doc Anson called it. Two-hundred-square-meter house, so about five hundred cubic meters—the house was twenty times larger than her customary quarters on-ship. It didn't make sense, but perhaps it was simply being groundside. She certainly felt more comfortable sitting in the tree and was thinking about heading back up into the branches when she heard the front door. Sandy was home, and it was time for them to talk.

"I don't trust him. I *really* don't trust *her*! If she were in my crew, I'd be asking Command to reassign her. Is there any way to convince Mom and Dad to disinvite those two?" Mace was doing one-handed pull-ups on the bar that Sandy had positioned in the kitchen doorway when he was thirteen. Their mother used it to hang clothes, but it was still in place and able to hold her now pretty much the same as it held Sandy in high school.

Sandy sautéed onions and mixed stuffing in silence for a moment before answering. "I don't think so. Disinviting someone once you've invited them is not the right way to handle this. It doesn't matter what we think of them. Dad was talking this morning and said that he credits 'Brother Eric' with giving him a 'new life.' Mom agrees. What gets me is that the timing doesn't fit. They met Eric and Elizabeth six months ago, but Dad started his rehab two years ago. Mom said last night that they started couples counseling *at least* a year ago."

"Six months ago? Really?"

"Yes. Something about being introduced at the Fellowship Memorial Day picnic."

"You realize The List was published May fifteenth? About two weeks prior?" Mace thought back to the day the crew roster for *Centauri Dreams* was published. She heard from Sandy for the first time in years, and her brother's correspondence with their mother led to this visit. It still didn't feel right. *Something was wrong.*

"Yeah, something is fishy, but today we have turkey!" Sandy stirred the last of the spices into the stuffing and put it in the oven to bake. "Now we just have to wait for everyone to get here."

The house smelled of turkey, side dishes, and pie when their parents and the couple from the Fellowship returned around 1300. Sandy planned to serve his Thanksgiving feast at 1400, so they would have to hurry to get the table set, drinks poured (fruit juice in deference to Dad) and serving dishes in place in time to receive the food as it came out of the various cooking appliances.

Finally, everything was ready and they sat to eat. Brother Eric was asked to say a blessing. It was rather bland albeit sprinkled with continued references to "purity of body" and "sacred creation." Talk throughout most of dinner was light. Mace and Sandy described their mission training, Sidney talked about his new job at the aerospace plant in Kerrville, and Melanie described the upcoming program for the symphony. Brother Eric didn't initiate any conversation topics, but he listened with interest, and asked questions throughout. Sister Elizabeth sat quietly through the meal despite Mace and her mother's attempts to draw her out.

The food was excellent, and the company was surprisingly relaxed. There were a few odd notes, though. One was that Sister Elizabeth would not take any dish passed to her by Sandy or

Mace. She had no problem passing dishes to and from Eric or the elder Wolfes; however, she demurred on any plate passed by the siblings, suggesting that they place it down on the table and she would "get to it in a moment." It wasn't offensive, just odd, although a couple of times Mace thought she caught a fleeting expression of distaste. The other odd note was Brother Eric's reluctance to talk about the Grace and Universal Peace Fellowship. Mace had tried to find out if it was organized as a church, and if so, what denomination it followed. Brother Eric dodged most questions and all Mace could glean was that the organization considered itself a social and service club, an extension of counseling support groups that were most members' introduction to the fellowship.

It wasn't until after dessert that the truth finally came out.

Sandy worried about serving food for company in the mismatched dinnerware—largely unused—belonging to his parents. Enough table settings for six people would have been a problem if Melanie Wolfe hadn't suggested they use the "Christmas China." It had been a wedding gift from her father who designed it for a pottery boutique. As a "family heirloom" it had seldom been used when Sandy and Mace were young, and it was a measure of how much their parents changed that it was even offered for this occasion. One of the most notable pieces of the collection were the "snowman" salt and pepper shakers. Short and round, they were all too easy to tip over and roll away—one of the reasons the set was not used with young children around.

Dinner and dessert were over and everyone rose from the table to make their way to more comfortable seating in the living room. As Sandy and Mace were picking up the dirty dishes to take them to the kitchen, the salt-shaker snowman was knocked over and rolled off the table. Mace was on the opposite side of the table and couldn't reach it, and Sandy's hands were full. Fortunately, Sandy was not dependent on just his hands. He lifted his left leg and neatly caught the knickknack with the tingers of his left foot.

Mel Wolfe gasped. Sidney actually laughed, while Brother Eric just sat with his mouth agape. It was Sister Elizabeth's reaction that broke the mood.

"Abomination!"

Everyone froze, but for Mace, suddenly it all clicked. "Grace and Unlimited Potential Foundation? Really? You're calling yourself that, now? Not 'Guardians of Unaltered Purity'?"

Sister Elizabeth's long silence finally broke, and she began to rant about abominations, gene-altered monsters, and killer cyborgs. Sidney and Mel just stood staring, but Brother Eric's own pretense slipped and he looked back at Mace with intense fury.

"Do you deny it? You can't, can you!" Mace glared back at him. "You damned GUPpies firebombed the clinic where two of my flight-school classmates were being treated after a shuttle crash. The investigation team determined that the crash was no accident. Then the damned GUP set the fire that finished the job!"

Brother Eric spat at her rather than answer. Sidney was asking for an explanation, but the representative of the rather infamous anti-human-augmentation group ignored him.

"Yeah, I get it now." Sandy returned from the kitchen with a dishtowel in his hands, still drying something. "It makes sense. The leader called himself 'Erebus' after the Greek personification of darkness."

Brother Eric turned toward Sandy in a menacing manner, but Sandy dropped one end of the dishtowel, revealing the foot-long carving knife they'd used to cut the turkey.

Brother Eric stopped. Mace was crowding Sister Elizabeth, trying to push her toward the door instead of the living room. The woman took a swing at Mace, but the latter simply held up her hand—her left hand—and blocked the punch. A soft "pop" sound suggested dislocation at least, if not a fracture of one or more small bones in Sister Elizabeth's hand.

Sidney surprised everyone by turning to the GUP couple and quietly but firmly saying: "I think you should go now."

The couple hurriedly left, Sister Elizabeth cradling her hand and alternating between grimaces of pain and expressions of anger. On his way out, Brother Eric swore he would file a criminal complaint against Mace for "aggravated assault."

Sandy washed the dishes and Mace dried them, carefully placing the Christmas China back into the cupboard. Sidney and Mel Wolfe sat quietly in the living room. Surprisingly, he was still drinking iced tea, while Mel poured just the slightest amount of brandy to settle her nerves.

Finally, Sidney spoke. "I guess I've failed you two again. I should have known that something was wrong by all the questions he asked about you two. I don't think they were even part of the local Fellowship before the news came out that you two were on the colony ship." There were tears in his eyes. "So now you'll probably hate me even more for this than for the mess I made of your childhood."

Mace moved over to where he sat, perched herself on the left arm of the chair, and leaned over to hug him with her right, biological, arm. "I'll admit that it's hard to forgive," she sniffed. "But I can't hate you for this. You were used."

Mel sobbed. "We didn't want it to be like this. We wanted you to have one good memory to take with you before you left forever!"

Sandy moved over to hug her, much as his sister hugged their father. He gave a short laugh. "Well, we have a memory, alright!"

The rest of the evening was punctuated with tears, laughter, and perhaps a small amount of healing. It was late when everyone went to sleep...except for Mace. She didn't need as much sleep since receiving her cybernetic implants. Pilots spent long hours jacked into the controls of their ships and shuttles, so one of her implants filtered fatigue toxins and metabolites to give her longer endurance. She went back to her retreat in the tree and stared at the stars. She still wanted to go. More than anything, she wanted the stars, but for the first time she was thinking of those left behind.

Forgive? Forget? Love? Hate? Those were abstract thoughts and all but meaningless outside relationships. Sure, Sandy was going, too, but instead of a three-year age difference, he would age an additional twenty years while she slept all the way to PCb. Her parents would be long gone by the time she woke up, and for once she found that it mattered to her.

It was well past midnight, and she was contemplating going inside when she noticed the shadow moving next to the house. There had been occasional movements inside the house, but those could easily be late-night bathroom calls. Outside the house was a difference matter. She dialed up the night vision in her cybernetic left eye and could see the figure clearly. Male, long stringy brown hair, straggly beard, bent over the gas meter. There was a faint reflection from his right hand, and Mace could see the type of handle used to turn the main gas supply valve. *Sandy mentioned a problem with the gas supply, but this was not good.*

Stealth was not one of her strong points, especially not from halfway up a tree. Still, she climbed down as quietly as possible. It must have sufficed, since the stranger continued to work at the gas meter and valve until she was nearly at ground level. He stood up, and she froze in position, ready to leap the last couple of feet to the ground and run at him, but he apparently finished the job since he put the valve handle away in a pocket and went to the back door.

He held a gun-shaped device next to the lock. It was an automatic lockpick. *Well, that settled the question of whether he was simply an after-hours utility repairman making a late-night holiday house call.* Mace smiled to herself. She could have saved him the trouble, though; the door was unlocked ever since she came outside... *two... three hours ago?*

Curious as to his intent, Mace decided not to follow him through the door, but decided instead to use the "other entrance" Sandy told her about years ago. Sandy had moved in to the utility room above the garage when he started high school. It was far enough from the other bedrooms that he could sneak out a window at night. He wouldn't be doing that climb anymore in Earth gravity, but Mace could certainly use it to sneak in. She might even do it without waking Sandy.

Jumping risked making a noise, especially if she hit a weak point in the wall or window frame. Instead, she just pulled herself up with her left arm, pulled out the loose security grille and belly-flopped through the open window.

"It's not nice to invite yourself in without announcing," came a voice from the direction of the bed. Her night vision was still active, and she saw her brother sitting on the side of the bed, putting on his gravity braces.

"Shhhh. Someone's in the house!"

"I know, sis. You're not the only one with night sight and hearing," he replied. He was in shorts and a T-shirt, the thin metal brace bare against strangely thin legs. "Let me go first, Optimus Prime. I'm quieter." Sandy was indeed quieter and was down the narrow steps to the laundry room before Mace finished testing the top step to make sure it wouldn't squeak under the weight of her cybernetic implants. In fact, he was so quiet he was able to sneak up unnoticed behind the intruder as the latter was working at the stove.

As soon as Mace entered the kitchen, she switched on the light. The stranger turned, squinting against the sudden light, and ran right into Sandy's fist. As the intruder rocked backward, Mace closed quickly and tapped the man behind the ear with the edge of her left hand. The man crumpled to the floor, unconscious, just as she heard the sound of breaking glass from the living room.

Sandy turned to her, wide-eyed, and yelled "Gas! Get out!" as he dashed toward the bedrooms. Mace saw a flicker of yellow-orange light and understood the situation immediately. The intruder at her feet turned on the "faulty" gas feed and was rigging an "accident" with the stove. The object through the window was probably a backup plan, an incendiary device of some sort, thrown by an accomplice when the lights came on.

Shipboard fire training came to the fore—there was an unconscious man at her feet. Even though he had set the blaze, there was still an obligation to evacuate him. Even if only to ensure that he stood trial. On the other hand, nothing said she had to be gentle. Cybernetically enhanced arm muscles easily lifted the immobile weight and her enhanced leg muscles took them to the front door in two leaps. She tossed the man out the door. He landed in a heap about ten feet from the front steps. There would probably be broken bones, but that was not her concern.

She heard more breaking glass—probably more incendiaries. This old tinderbox of a house was going up fast. "Sandy! Mom and Dad!"

Sandy's voice called from the back of the house. "I've got them! Sending Dad out to you, I'm going to carry Mom." Smoke was billowing and Mace could see a figure coming toward her, bent over, and holding something against his face. It was her father, holding a scrap of Sandy's T-shirt over his mouth and nose. Mace guided him out the door and sat him down on the grass beside the still body of the man she threw out the front door. She touched a finger to his throat and received a full medical readout through the cybernetic sensors. Alive but unconscious. Her father was coughing, but a quick touch showed that he, too, was okay.

Where was Sandy?

Mace tried to reenter the house but was blocked by fresh tongues of fire. Sandy and Mom were trapped! Sandy was tough; he was bioengineered for adverse space conditions. His toughened skin and tolerance to low-oxygen conditions would even resist the fire, but he needed a way to get their mother out!

She heard a shout and more breaking glass. The bedroom window! She would have to pry out the security bars, but that wouldn't be a problem. She ran around the house to the windows outside her mother's bedroom. The glass was already broken out and Sandy was inside, trying to pry at the bars.

"Stand back, bro," Mace called, then jumped up, grabbed the base of the grille with her left hand, and planted her feet firmly on the brick outer wall of the house, the bulk of her body parallel to the ground. With a grunt, she applied force with her enhanced leg and arm, ripped the grille free of the wall and flung it behind her. There was a sound as the grille seemed to hit a soft object. She fell to the ground once the grille was free, but quickly stood up to take her mother, wrapped in a sheet, from her brother. Seeing that Sandy was able to climb out unassisted, she started to unwrap the sheet from her mother.

"Careful, she was hit by glass." Sandy put a hand on his sister's shoulder before she could finish. Mace could see a dark stain on the linen.

It was blood.

"Let's get her into the light." Mace stood, delicately lifted her mother, carried her to the front yard, and laid her on the grass next to her father.

"Mel! Melanie! What happened?"

"It's okay, Dad. We've got her, but I need more light."

Sandy shrugged, gesturing toward the burning house. "My stuff's in there; a bonfire's not good enough for you?"

"I'm pretty sure Goon One over there has a flashlight. Grab it."

Sandy pulled a flashlight from the arsonist and shined it on their mother. There were several small splinters of glass embedded in her skin, but one large one was embedded in the neck. The blood coming out around the sliver was pulsing. "Damn. It might be in her carotid. Don't move it."

"No, this has to come out. And it needs to be sealed or it will cut further and she'll bleed out." She bent over for a closer look while motioning for Sandy to hold back their father. Before she took the flashlight, though, she started to remove the skin-like covering over her left hand, noted that it was torn and melted, and stopped. Her index finger was exposed, its blue crystalline surface and faint lights now clearly visible. She bent her thumb over the palm, then curled her other three fingers over it. A faint whirring sounded and small metallic probes extended from the fingertip.

"Hold the flashlight right...there." She leaned forward and dialed up the magnification in her left eye. With her right hand, she grasped the piece of glass and placed her left index fingertip at its base, the probes entering the cut. "One...two...three..." She pulled out the glass and jammed her finger into the wound. There was a flash of light, and the faint smell of burnt meat. "That should hold her for now."

The house *popped* and embers flew out. They could hear sirens and saw flashing lights in the distance. Another pop, another spray of sparks, and the bedroom end of the house collapsed.

There was a scream from the back yard.

Sandy and Mace looked at each other. "Goon Two," they said simultaneously as Sandy hurried around to the source of the scream.

Mace grabbed her father's wrist and pulled his hand over to keep pressure on her mother's still-seeping neck wound. It wasn't pumping blood, but pressure needed to be maintained for several more minutes. She headed after her brother as the first emergency vehicles arrived—police and fire trucks. She found her brother standing over a man on the ground, his legs trapped under the mangled mess of the security grille she carelessly tossed behind her during the rescue.

There were spots of fire now burning in the grass, and in a leaf pile about a foot away from the recumbent man. It was Brother Eric. Sandy was standing over him, one foot on his chest, fingers knotted in the collar of his shirt.

"What, you didn't expect me to be strong? You thought because I didn't meet your expectation of a human—that I was bioengineered for low gravity—that I would be weak?" Sandy stood over the intruder, body language signaling anger and rage. "You argue about biological purity, about 'unaltered' humans, yet you live with modern medicine, vaccines, gene therapies and corrective surgeries."

"He dyes his hair and obviously likes cosmetic surgery just fine," Mace said.

"What?" Sandy looked over at Mace, who merely tapped the side of her head beside the left eye.

"Eh, simple spectrography," Mace said, dismissively. "Diamino-toluene in the hair means hair color. Probably to cover the gray and change his appearance. Fine scars around the nose and eyes from plastic surgery—either vanity or to fool facial recognition. There's a scleral scar and artificial lens in his right eye."

Sandy practically snarled. "So, correcting your vision and changing your appearance with surgery is okay for you—just not for the people who are trying give mankind a future?"

"Not 'people.' Monsters." Brother Eric tried to spit in Sandy's face, but given that he was now being held on the ground by both Mace and Sandy, the spittle just ended up dribbling from his cheek.

Sandy pulled back a fist to hit the fake minister, but Mace laid her flesh-and-blood hand on her brother's forearm to prevent the gesture. She shook her head, sadly, and reached out her artificial hand to grasp Erebus firmly by the jaw.

"A few pounds of pressure, and you'll have a nasty bruise; ten pounds of pressure and your jaw will dislocate. A few more pounds, and I can break your jaw, but then again, Sandy can do that with his fist, or my father could do it with a steel bar. Fifty pounds of pressure, from my fingers alone, and I could so thoroughly crush your jaw that the doctors could never hope to rebuild it—you'd never speak again, nor chew. You'd take all of your meals through a straw." Mace released her grip and she could see the white, bloodless patches from her fingertips. She patted him on the cheek, none too gently. "On the other hand, it probably wouldn't silence you, would it?"

She looked up at the approaching police officer. Sandy and Mace both stood up, and Mace grabbed Erebus by the elbows and pulled him to his feet. She tightened her grip just enough to pinch the nerves so that he wouldn't fight the handcuffs.

"It's not our bodies that make us human, Mr. Erebus. The secret is choice. I choose not to take your life, an eye for an eye, a tooth for a tooth. Not because you are human, Mr. Erebus... but because *I* am!"

Sandy believed that the people who designed hospital rooms must have had a private glimpse of Hell. Either that, or Limbo. Drab green walls, hushed background noises, and the steady *beep-beep-beep* of the heart monitor would lull anyone other than the patient to sleep . . . or catatonia. Mace had apparently succumbed; she was racked out on the convertible sofa. Their father was reading in the tall-backed reclining chair for ambulatory patients, but his eyes were now closed, and his head was back against the chair. Sandy was seated, simply holding his mother's hand and waiting patiently.

Mel Wolfe coughed, and her voice was hoarse. "Sidney? Sandy? Mace?"

All three were instantly alert, rose, and clustered around the bed. Mace poured some ice water into a cup, placed a straw in it, and held the straw to her mother's lips. After a few sips, Mel began again. "Sandy and Mace. I'm so sorry. We had no idea what that horrible man planned."

"Actually, I do now." Sidney Wolfe held his head high and looked his son and daughter in the eye. "He had grenades labeled with 'PCbCorp.' He wanted to provoke an 'accident' and make it look like you two were at fault." He looked down at the floor. "He wanted an incident to discredit the colony mission and everything it stands for."

"You don't know how much I envy you two," Mel said. "To finally be brought back together, and now you're leaving . . . forever!"

Sandy and Mace shared a look.

Sandy cleared his throat. "Actually, it doesn't have to be forever. If you're serious, the colony needs engineers *and* artists. Not this trip, but Mission Two leaves in ten years, and Mission Three five years after that."

"They won't want me. I'm old and a broken-down drunk."

"Actually, Dad, you're as healthy as we are." Mace spoke up. "When I checked you after pulling you out of the house, I got a nanocyte read-back. When you decided to dry out, the hospital ran a full nanoscrub on you. You're as healthy as a twenty-year-old, with about the same additional lifespan." Sidney's face brightened, and Mel's expression held a glimmer of hope. "You too, Mom. I spoke to the doctors here, and the risk of reopening the carotid meant they gave you nanocytes, too."

Sandy said what they were all thinking: "We'll be a few years older, but maybe it's time for a fresh start!"

Mace knew it was true. *Forgive, forget, and start over. It was the human thing to do.*

Securing the Stars

The Security Implications of Human Culture During Interstellar Flight

Mike Massa

Mike Massa has lived an adventurous life, including stints as a Navy SEAL officer, an investment banker, and a technologist. Newly published by Baen Books, he is coauthor on books five and six of the Black Tide Rising apocalyptic thrillers with NYT bestseller John Ringo. When Mike isn't writing science fiction, he keeps his hand in as a cyber-security researcher, consulted by governments, Fortune 500 companies, and high net worth families on issues of privacy, resilience, and disaster recovery.

It is the year 2150 and we can finally dispatch the first human interstellar voyage! Skycorp—whose orbital foundries and building yards have been enabled by extreme twenty-first-century drops in the cost to reach high orbitals and L5—has completed the first ship capable of prolonged duration, albeit, slower-than-light (SLT) spaceflight. Intended to reach a rocky exoplanet that has been confirmed to orbit plumb in the middle of the Goldilocks Zone of a G2-type star, only twenty or so light-years away, the ship is ready for the one-way trip.

It is now just a matter of packing in the pilots, mission specialists and passengers, and then off we go!

Right?

Well, maybe. The first successful interstellar mission will owe its success to the culture that humans bring with them for the long trip as it does to the enabling technologies. So, how must space-bound humans evolve their culture for a small group to survive not only the trip, but each other, and ultimately reach their destination to achieve the even longer-term goal of permanent human settlement outside our home star system?

The technological requirements of the voyage alone are daunting. Until the ship attains a stable orbit above the new planet, every foreseeable item that the crew might possibly need (or the means to make it) must be included at the start, decades or centuries before they arrive. The life support, from enclosing the crew inside a humid and warm gaseous envelope, to the process of recycling food and water efficiently, must be maintained in perfect order without interruption. The propulsion and navigation systems, ditto.

For this article, let us assume we have surmounted every terrestrial technical challenge presented to our race thus far—the engineering challenges of interstellar flight. The remaining persistent challenge that we face on Earth is the same one we will inevitably import into the spacecraft—us. The human element represents the most challenging component of the mission, which must not only be secured, but *assured* as well.

The resilience of our current biosphere creates a buffer that prevents even massive catastrophes from destroying human civilization. This is true because of the vast size of the shared space on our planet as well as the robust supply of needed environmental inputs (breathable gas, water, food, etc.). The planet and its ecological systems are inherently resilient even to extreme shocks. This is precisely the opposite of the environment aboard an interstellar spacecraft.

Humans are inherently messy, in both a literal and figurative sense. The vagaries of human behavior will introduce variables that must be considered in advance, much like the supply chain that will provision the spacecraft. This is especially true when the shared living and work space of the crew is even more fragile than that of a commercial jetliner. Whether by design, negligence, or an exceptionally unlucky set of circumstances, a single human can put the entire interstellar mission at catastrophic risk.

Today, humans routinely transport themselves in fragile,

technologically advanced aircraft: high-speed, high-altitude passenger planes. To safely reach our destination, we accept extreme incursions into privacy, from electronic imaging, which publicly strips away our clothing, to invasive searches of personal property. Furthermore, passengers and crew have demonstrated their resolve to act in concert, spontaneously, to address individuals that break the shared behavioral compact during a flight. They do this despite the usual ground-side inhibitions against such action because they understand the intrinsically fragile nature of aircraft; therefore, they must temporarily adopt their "travel culture." How much more adjustment is needed on a spacecraft? Note, the lethal combination of effective vacuum and near absolute zero temperature of the medium traversed by our notional starship makes the conditions outside a modern jetliner during a Los Angeles to New York City hop seem merely dangerous.

Recognizing the precarious nature of an artificially sustained environment needed by humans during interstellar flight, we can consider the amount of freedoms that may be allowed to the crew and passengers of the spacecraft.

However, before we can stipulate how humans "must be" to complete an interstellar voyage, we must first identify the risks during such a trip. The greater the specificity that can be achieved, the more confident we can be when describing what we—the planners, designers, and crew of the trip—will need to address before ever leaving Earth's orbit. Yet, why should we man the ship at all if the environment is as perilous as stated above, or if humans are as dangerously messy as presumed?

There are ongoing exploratory missions in our own solar system. These unmanned efforts demonstrate the degree to which autonomous and semiautonomous spacecraft and landers can operate in hostile environments. That autonomy doesn't scale well against time and is likely an artifact of the early stage of autonomous-systems engineering in which we find ourselves. Depending on the platform, human supervision and intervention is currently required at intervals ranging from minutes to hours. To assure the success of unmanned systems, especially where human safety is paramount, this interval drops to seconds or continuous, real-time supervision. One example is the employment of lethal force from military systems where our aversion to machines independently making life-or-death determinations is so strong that there are international agreements in place to prevent the deployment of armed autonomous systems. If we were to accept that the ship will carry humans in some form of hibernation, then we would have to accept that machines will be making entirely unsupervised decisions about health and safety for extended periods. This is an example of cultural norms driving applications of technology.

By 2150, the state of the art will have advanced tremendously. Even if we stipulate the advancements in software engineering, artificial intelligence, and autonomous systems make an entirely unsupervised voyage possible. Therefore, it is important to understand the core mission of the colony effort. That objective will be to reduce the existential threat to our race by creating a self-sustained human presence outside our solar system, and possibly outside our immediate neighborhood, which is the spiral arm of the galaxy. Ultimately, our notional interstellar spacecraft *will* carry humans.

Therefore, while technical advances in the era leading to human interstellar spaceflight may support unmanned probes, or perhaps support wholly unsupervised spacecraft operation where humans are in some form of suspended animation, it is unlikely that terrestrial culture will allow ship operators to cede human welfare solely to machine intelligence. In this event, we will have to build a ship and a culture that will operate under the direct control and supervision of alert, awake human adults.

Our space-bound humans will bring with them a culture composed, in part, of the dominant culture in their nation state of origin. They will also bring cultural aspects from the techno-scientific community from which they are sourced and, perhaps, from the selection and training environment that produces the crew and passenger cohort.

The successful crew will have a culture that addresses the physical and emotional challenges of long-duration spaceflight. This culture is a framework of mutually expected and accepted behaviors, permissions, and values. Because of the radically different living environment in space, no land-based cultures are likely to be a good match for the needs of the voyagers. This is especially true for the risk-management part of culture, which can be labeled "security."

What is the nearest Earthbound analogue to the possible rigors of multiyear deep spaceflight? In which human endeavors do we isolate small groups of persons for long periods, often in hostile environments, and expect them to perform repetitious and tedious, yet highly technical, tasks with great precision? How do these groups operate without the support of a family structure or the freedom to change their circumstances if boredom or danger grow unbearable?

There are only a few current possibilities that begin to approach the combination of isolation, tedium, danger, and duration, which will characterize interstellar STL travel: nuclear deterrence submarine patrols and over-winter Antarctic stations are two that come to mind. Historically, long-duration sail voyages of exploration also share some similarities with long-duration spaceflight. All are but pale comparisons to the rigors of multiyear spaceflight.

Table 1 offers a comparison of the candidate cultures. The vertical columns represent the cultural and environmental variables which may predict security-culture risk for spaceflight.

Some interesting implications for security culture may be inferred from these examples, disparate as they are.

The mission duration can be considered both the risk window (how long the risks have to manifest) before the ship reaches its

TABLE 1: Comparison of voyage factors

LOCATION	EXAMPLE	DURATION OF DUTY	CREW SIZE	LIFE EXPECTANCY WITHOUT HABITAT*	COMMUNICATION WITH POINT OF ORIGIN	PLATFORM RESILIENCE
Deterrence patrol	US nuclear ballistic missile submarine (20th century)	60 days	120	Seconds	Weekly	Low
Antarctic	South Pole Station (present day)	8 months	70	Up to 1 hour @ −60°C	Daily	High
Trans-oceanic sailing	European ship (18th century)	10 months– years	50–100	2–4 hours @ 10°C	Zero	Moderate
Interstellar voyage	Spacecraft (22nd century)	20+ years	160	Zero	At start: hourly By mission end: impossible	Very low

*Assuming no replacement miniature habitat

destination, as well as a factor that can lead to greater security risk because of morale issues. Longer-duration missions mean more time for something to go wrong and more opportunities for the crew to be negatively affected by their environment.

The crew-size estimates in Table 1 are not intended to be the basis for a discussion about what the minimum sustainable crew size may be, or what amount of unique genetic patterns is needed for perpetuating the race; rather, it is intended to provide a sense of how many potential, deliberate, or accidental, sources of failure points exist in each platform. In this sense, each individual represents a decider who may make good or bad security decisions, be they intentional or unwitting. Therefore, larger crews may equal greater risk.

The relative danger of the operating environments is illustrated by the crew-survival estimates outside their host platform. As survival probability lowers, stresses on cultural standards will increase.

The frequency of communications with the point of origin represents both the ability of the crew to "reach back" for technical consultation as well as individual crew member morale support. Thus, to reach back is a way to renew or sustain culture.

Lastly, the resilience of the platform is an indicator of how robust the systems are when exposed to shock. Relatively fragile physical and logical constructs, such as aircraft or flat networks, can withstand much smaller shocks compared to functional analogues, such as buses or highly segmented networks. The subjective grade of each platform in the table illustrates the resilience during operating conditions. For example, a submarine on the surface with ample reserve buoyancy is far more resilient to collision than a submarine operating a few boat-lengths under the surface. South Pole Station is more resistant to shock in the high summer of mid-December than during the three-month period of total darkness and windchill temperatures below 100 degrees centigrade in July. Our notional spacecraft is more resilient when in space dock at L5, surrounded by contractors and heavy equipment, than it is several light-years beyond Sol's heliopause. The comparative resilience of the platform is a driver for different cultural norms.

One of the most interesting questions is "how much security is enough?" A strictly military dictatorship might have the most kinetically shockproof culture but can also be viewed as oppressive. Instead, it may be useful to correlate the risk set for each example with the minimum amount of security needed to guarantee survival for the entire platform. Where risks are high, and resilience is low, we can expect security considerations to be higher, and restrictions on personal freedoms and privacy to be correspondingly high. This is consistent with what we can observe to be the case historically on sailing ships as well as on twentieth-century strategic deterrence submarines or the "winter-over" crews on Antarctica. Where the margin of resilience is relatively low, the strictures placed upon individuals are relatively high.

The security-oriented culture for these platforms always depends upon the physical presence of trusted personnel, often in multiples, in locations that house the equipment, which ensures the immediate safety and security of the platform. On a sailing ship, these locations were three: the powder magazine, the ready fresh water supply (called the scuttlebutt), and the captain's person (for navigation). A single individual could terminate a mission by acting decisively at any of these three points.

Each point represents a critical path node, without which the platform may not only be unable to complete its mission, and whose disruption can lead to the loss of the platform.

On a twentieth-century submarine carrying nuclear missiles, there was a similar calculus. The nuclear powerplant, the communications center, the launcher control center, the missile command center, and the conning deck (from where the sub was commanded) were all manned continuously. Some of the more sensitive spaces included armed watch standers. Again, the heuristic that led to the protection of these spaces above all others was the recognition that the ship's mission and physical integrity could be catastrophically disrupted from these nodes.

The crew members of both seventeenth- and twentieth-century ships accepted considerable constraints on the amount of personal freedom that they enjoyed. The maritime discipline was imposed because of the element of shared danger, mutual interdependence, and sense of mission. These crews lived in a culture where one watched out for fellow crewmates, not just because they were comrades, but because a single actor held the safety of all the rest in their hands. This culture of interdependence and security was immensely foreign to land-based visitors.

Even the very language used by sailors diverged from sho-reside lingo. The reasons for a unique lexicon was practical and based on the security culture. Being able to precisely lay hands on a piece of equipment or perform a technical task in the dark, during moments of extreme stress, was literally a life-or-death capability. What a landlubber regarded as a quaint affectation, was actually a material expression of the specialized culture. On land, we simply connect ropes together. At sea, a sailor may bend two lines together, tie off a hawser, or quickly make a halyard. Each of those are quite distinct and in a worst-case scenario, confusing them could put the ship at risk.

In comparison to systems of ropes, cordage, and wood, our spacecraft is terrifyingly complex and fragile. A specialized framework that assures its safe operation will be needed. Earlier, we recognized that even untrained, but heavily socialized, passengers on commercial aircraft were sufficiently acculturated to the modified social compact present aloft to adopt new security behaviors. Meanwhile, in-flight persons have proven themselves capable of recognizing security issues and self-organizing to restrain dangerous actors.

We have also begun to address specialized language as a part of the security-aware culture on ocean-going vessels. This long-term behavior change permits more precise and efficient communications during an emergency afloat.

Spontaneous security cooperation and specialized language are two parts of a security-culture framework which will be needed during long-duration spaceflight.

Business consultants delight in the opportunity to delve into "risk management frameworks"—a semi-arcane specialty with its own language and correspondingly high bill rates. Thus, we can use much simpler language to think about how our space-bound framework will need to be supported by a crew culture that voluntarily lives and breathes the principles that protect the ship and crew from all threats, while relinquishing some highly valued aspects of terrestrial culture.

However, for reference, there are several different ways of organizing a set of rules, or governance, to think about and manage the different organizational risks that already exist. An exhaustive categorization of the various types of security is a useful resource to supplement the consideration of security-aware

TABLE 2: Representative risk frameworks

OWNER / PUBLISHER	FRAMEWORK NAME	SPECIALTY / TYPE
National Institute of Standards and Technology (NIST)	Special Publication (SP) 800 series	Comprehensive catalog addressing physical and information technology risk, security, and controls
International Organization for Standardization (ISO)	ISO 31000 (also see ISO 20000, 24762)	For-fee risk management with a functionally global span, from personnel to physical plan to systems and controls
Information Systems Audit and Control Association (ISACA)	Control Objectives for Information and Related Technologies (COBIT)	For-fee business-centric framework, spanning processes, controls, management, and a maturity model
National Fire Protection Association (NFPA)	NFPA 1600 series	Standard on Disaster/ Emergency Management and Business Continuity programs

space culture. Terrestrial specialties include information security, supply chain security, personnel security, etc. Table 2* lists some, but by no means all, examples.

Rather than think about our crew culture only in the context of a risk framework that addresses things like information security, physical security, etc., it is more interesting to think about the stages of the mission as well as the cultural baggage that our crew will inevitably bring.

There are three basic phases to the voyage where threat vectors may be characterized differently: Earth-orbit training and construction, near-Earth operations and departure, and extrasolar operations. In the first category, we can stipulate that the security culture of the ship will be overshadowed by extensive, direct support from Earth. In the second category, the ship will be physically separate from Earth, but close enough to permit both communications and even physical intercept. While Earth security culture will still be relevant, the process of the crew detaching from Earth will be well underway. The second phase will be relatively short.

The third category is the focus of this thought experiment. While an open-space intercept from extraterrestrial forces is

---

* The author does not endorse any of these options; the list is for information only.

possible, the likelihood that the engagement features a civilization so closely matched in the level of technology with Earth that our defenses are relevant is vanishingly low; therefore, this scenario is omitted from consideration.

That leaves four fundamental risks whose point of origin may be identified: spontaneous ship-system failure, collision with a celestial object, unwitting human error, or deliberate crew or passenger actions. Our spacers' culture must be optimized to address these four categories of risk to their mission and themselves. What sorts of behaviors will our crew want to adopt and reinforce? Which Earth-centric behaviors and expectations must be adjusted, or even abandoned, so that these risks might be minimized?

Principle areas of human behavior that drive culture on Earth will have to be adjusted. Some will change naturally in the absence of external drivers. One such example is the set of behaviors and competitions arising from consumer-driven mercantile activity and class-segregated consumption. Some will be persistent, such as behaviors arising from human sexual drive, while others may be exacerbated, such as behaviors arising from expectations of privacy. The mechanisms for conflict resolution will need adjustment to accommodate the realities of constrained shared spaces, long-term proximity to crewmates, and potentially fragile ship systems that aren't resistant to violence.

Because the consequences of conflict are so severe on a space voyage, the means to avoid social conflict via consensus-driven problem solving is critical. However, because critical systems are relatively accessible, even one disaffected crew member could cause catastrophe. The social safety net, which must sense both internal and external conflicts among the crew, will sharply erode the ability to maintain anything resembling the privacy we enjoy on Earth. In essence, anyone's business becomes *everybody's* business, and thus will necessitate thorough monitoring of behavior patterns. Deviations from established personal shipboard routines must be autonomously detected and human leaders alerted in advance of actual events.

Teamwork is a hallmark of the existing space programs. The stories of the bonds formed between crew members in the Apollo, Space Shuttle, and International Space Station programs are legend. In a fixed-crew complement, the specialization is inevitable, but duplication in skill sets will be important for crew resiliency. Cross-training increases the likelihood that the right people with the right skills are immediately on hand during a localized emergency.

Because of the strong degree of interdependency among the crew, personal respect and accountability will be needed to exceed Earth norms. Examples include expectations regarding minimum levels of fitness, maintenance of diverse and demanding skills sets, dietary practices, and most incendiary of all, religion.

The confrontations possible between incongruent religious viewpoints can generate hostility and violence. Moreover, nearly all religions can spawn fanaticism and extremist behaviors. There are religious conflicts whose origins stretch back more than a millennium—and the impact on individuals in the affected populations have correspondingly profound cultural effects. The large-scale migration of disparate populations into western Europe during the period 2015–2017 and the ensuing unrest and political consequences are an example. Compacting this dynamic into a very small, unescapable habitat will result in undesirable outcomes. This predictable reality means that selection of spacecraft crew and passengers must screen out individuals who cannot subordinate religious priorities to the security culture critical to the spaceflight scenario outlined throughout this article.

Recalling the earlier statement that humans are inherently messy, we can predict that, despite all precautions and efforts to reinforce ship culture, some corrective measures beyond group dynamics and reinforcement will be needed. Requirements will range from denial of routine privileges to physical confinement. The shipboard culture will have to be flexible enough to recognize the need for rules that address behaviors outside the boundaries of what is acceptable.

Crew flexibility must include acceptance of extreme protective measures, when necessary. This is another example where space culture must diverge from Earth culture. The traditional way to resolve intractable problems on Earth is to create a political system that allows humans to make fundamental changes to their social construct, often in response to changes in their environment, or access to resources.

Since the voyage will feature relatively static conditions, big changes to the social construct during the long, yet finite, voyage may be counterproductive. This may require the biggest cultural shift of all: Political organizations and sectarian divisions among the crew must be rejected for the duration of the voyage. This could be particularly hard for Westerners to accept given their political tradition of self-determination and individualism.

Table 3 considers selected elements of terrestrial culture, summarizing issues and offering adjustment vectors for each cultural norm touched upon in the foregoing discussion.

In short, if we use humans and not machines to shepherd our race to the nearest stars, then we will do so by embracing the risks of using messy, unpredictable, and fallible humans. However, those spaceship crews will have to drastically adjust their cultures to optimize the likelihood of safely reaching their destination. The adjustments to familiar cultural norms around relationships, politics, reproductive activity (even of the recreational variety),

TABLE 3: Cultural norms

CULTURAL COMPONENT	TERRESTRIAL NORM	SPACE NORM	ADJUSTMENT VECTOR
Conflict resolution	Self-guided resolution as an option for antagonistic pairs; violence possible.	Group guided resolution; violence not permitted.	Decreased scope of personal action
Sex	Hidden, not freely discussed; serial monogamy	Unavoidable awareness; pairing shifts unpredictable	More liberal
Religion	Theocracy present, but distribution of religious impact highly variable.	Religious toleration; secular, technical, and social constructs	More secular
Privacy	Accepted as a fundamental right	Highly prized, but profoundly limited	Less liberal
Language	Freedom of language use endemic	Requirement of mandatory language standards	Less liberal
Discipline	Highly variable; generally incremental	Very strict; few gradations	Less liberal
Fitness / education	Self-guided, optional	Required, testable	Much less liberal
Political organization	Highly variable, but mostly liberal	Impermissible	Totalitarian
Teamwork	Optional; personal preference	Mandatory; spontaneous self-organization	Less liberal

discipline, and language will be significant and mandatory. The changes would be reasonably viewed as totalitarian and unacceptable in nearly any other environment. In total, these adjustments will form a new, space-based cultural framework that will seem quite foreign to ground-siders...

...but that framework will become the daily environment for *Homo stellaris, or the humans who live between the stars.*

The Smallest of Things

Catherine L. Smith

A science fiction reader from an early age, Cathe Smith grew up wondering about alien planets. Since she couldn't study aliens in college, she studied insects, which most people would argue is close enough. With a background in insect molecular genetics and evolutionary systematics, she has been a consultant to numerous science fiction authors looking to create interesting aliens. After doing that often enough, she decided to try her hand at writing her own stories.

There are a few greetings I have received when walking into my lab that woke me up faster than the first cup of coffee:

"The lab is on fire."

"The ceiling collapsed."

"The good news is no one died."

This time, my lab manager, Dr. Tara Hauff, went for a subtler approach. "Doc Morgan, take a look at this data—it looks...weird."

"Which data?" I asked wearily. We had collected so much of it since the ship touched down that I was starting to see it in my dreams.

Tara walked over to the main computer consoles for the lab as she continued. "You know how we noticed some small anomalies in the orbital scans? I think you were right about throwing out our assumptions and working from the ground up."

She spoke as she pointed toward a split screen displaying two different chemical readouts. "The one on the right is from

239

some lettuce in the greenhouse, showing the standard four peaks indicating the four DNA bases of Earth. The one on the left is from one of the vegetation samples collected outside the glass yesterday."

The graph she pointed to had six peaks.

"Does this indicate what I think it does?" I asked. "That we're dealing with two additional DNA bases—alien ones at that?" My brain whirled, no coffee required.

"I think so," she answered. "I've run the tests several times on different samples of the native Thorbian vegetation. I reset the machines and had the drones use new reagents and everything. It all comes back with six peaks."

This was huge, if we were right. Identifying life forms on another planet that used foundational blocks similar to our own would be the scientific discovery of a lifetime.

"Run it again, Tara. Sanitize and sterilize everything. We must be certain. Grab new vegetation samples and see if the light traps outside the glass caught any small animals that we can test. Let me know if you find anything else unusual." I wasn't much of a morning person, but I was truly awake now. I turned and headed for my office.

The *Virginia Dare*—the Starpherra Pharmaceutical Corp. ship that brought us to our new home—had been orbiting Thorbia for the past six local months, running analyses to verify the long-range data that had classified it as Earthlike. Our new home was flush with potentially exciting and profitable discoveries. It was like catnip to a scientist and irresistible to the colony's financial backers. It didn't hurt that Thorbia was breathtaking in its splendor with a vast unspoiled alien wilderness.

My lab was part of the advance survey party which made planetfall while the ship remained in orbit. Captain Eugenia Richardson, skipper of the *Ginny D* and interim commander of the expedition, joined us on the surface to personally oversee site preparations. While on the ground, our mission was to determine the habitability of Thorbia before the bulk of the colonists were roused from stasis. If everything came back positive, then the ship would begin dropping cargo modules. Each would make a one-way, controlled entry into the atmosphere and land at the site of the new colony. After that, there was no turning back.

I entered my tiny office and opened my email. There were the usual queries about projects and requests for everyone to wait patiently when requesting materials from Inventory. Thanks to Tara's news, I would soon need to requisition live lab animals for some of my experiments. Rats and rabbits to start; eventually pigs and then monkeys for full biocompatibility tests.

Weird DNA notwithstanding, I didn't want anyone to sicken and die from an unknown alien disease; we were still studying how terrestrial microorganisms interacted with alien biological material. We'd instituted full quarantine protocol: We used autonomous remotes for everything we could and no one went outside The Glass unless they were in Class B protective gear or higher. Even with all these precautions, I feared it would only be a matter of time before someone was exposed.

A few hours later, an excited Tara poked her head into my office. "Molly, there were some small flying things caught in the light trap that I could use, so I reran everything as you suggested. Those extra bases are real. This will get their attention at the next Captain's Lunch."

"Great work—I knew you were the right person for this stage of the expedition. Get some of the techs analyzing the amino acids and subsequent protein structures. We need to know if or how these differences are going to affect Terran compatibility with Thorbia," I instructed.

I pulled up the data on my screen and studied the peaks and valleys. Such small things to have such large implications for our future.

Later that evening I heard a familiar pattern to the knock on the door to my private quarters. Sliding open the door, a towering man engulfed me in a warm bear hug before kissing the top of my head. I felt myself melt into his embrace before releasing him and beaming up at his face.

"I missed that smile. Have a good day in the lab, my love?" Alan Jukka and I had begun dating before the ship departed from Earth and I was thrilled when he decided to come to Thorbia with me. His intense green eyes initially attracted me, but I fell in love with his heart and sense of humor.

"I love being on an alien world. There's so much cool stuff to discover," I gushed, dragging him further into my quarters. Even as a security lead, Alan only had a bed in the barracks, so whenever possible he came to my quarters for some much-needed privacy and relaxation.

"What new thing did you discover today?" he asked, smiling.

"Thorbian life has extra genetic bases," I burbled excitedly. "This was something they only theorized about on Earth, and here we've found it!"

"Extra genetic bases." Alan didn't always understand the details of what I did, but he loved my enthusiasm for it.

"Okay, so you know how on Earth we have four genetic bases that make up our DNA? A, T, C, and G?" I asked.

"Uh, sure," he cautiously answered.

"Right, so Thorbia has A, T, C, and G, but it also has two completely new bases that for now I'm just calling X and Y."

"Is that good? Or bad?"

"Beats me. I have my people working on what this means for amino acid coding and what effect this will have on the proteins formed, how they fold, and everything that comes from that." We migrated to my bed and I was now in my favorite spot, nestled into his arms.

"Amino whatsis? Do you fold proteins like you fold laundry?" he joked.

"Amino acids. On Earth, three DNA bases in a row like AAA, ATG, CCG, and so on signal for one of about twenty amino acids. A hundred or so of those, and you have a gene—basically a blueprint for what amino acids are required to be linked together in a chain to make a protein. Once the chain is complete, the protein starts to fold into a complex 3-D shape that our body will use. These things are tightly controlled. The wrong-shaped protein can have lethal effects."

"I think I get it. So how will these extra bases you found change things?" Alan asked.

"I don't know. No one does. I have everyone I can spare in the lab trying to figure it out. Tara was going to stay late to keep working on it, but I told her to take a break and go spend some quality time with her husband. Enough about me, though. How was *your* day?"

"I don't mind, babe. I love your excitement about this." I felt his light chuckle as it rumbled through his body. "Work was okay, standard patrols to make sure protocols were followed, stepping in to deescalate arguments between people when Inventory tells them they have to wait their turn. For a bunch of smart people, the concept of mission priority seems to be completely beyond their comprehension. There's this one dude from the Physics section who has a talent for being an ass. Today, he made one of the desk techs cry."

"That sounds like something one of Mark's people would do. I'm sorry you have to deal with grown adults acting like petulant children." I turned to kiss Alan. If his day sucked, I could at least make his night a little better.

The next few weeks were a blur of countless tests and frantically reprogramming everything to account for the new bases. Part of my lab worked on teasing apart protein structures and blueprints from what was collected outside The Glass, while others started deciphering the amino acid codes that built the proteins.

Like the landscape of Thorbia itself, what emerged was tantalizingly familiar yet subtly different. Of the twenty-five amino acids we found locally, twenty were identical to the standard Terran ones, but five were unique to Thorbia. Fifteen of the common amino acids showed up in roughly the same proportions on both worlds. It appeared that the unique Thorbian ones had partially taken over those roles. This difference in amino acids created Thorbian proteins that would perform the same functions as proteins on Earth, but they did so using a slightly different structure.

The time was rapidly approaching when I would need to move from computer modeling to live testing. I pulled up my emails to check on my requisition and was pleasantly surprised to find that it would be ready a few days early.

"Tara, let's get the quarantine cages ready for occupation. The rats and rabbits will be ready for pickup in two days."

Tara nodded. "Got it, Dr. Morgan. I'll get someone to run diagnostics on the manipulator arms, too."

"Thanks, Tara." My curiosity and excitement burned inside me. *What would Thorbia hold in store for humanity?*

The meeting-room lights were dimmed for this portion of the Captain's Lunch so everyone could see the results of my lab's trials.

I controlled the customary exuberance in my voice as I narrated. "As you can see, ingestion of the native vegetation causes massive anaphylaxis in Terran rabbits." On the screen, the albino rabbit fell over on its side, heaving as its airways closed, paws twitching in futile panic. I had muted the videos—the images were enough. Adding audio would have been too much for those on the ship unused to the horrible cries of a dying rabbit.

I continued narrating over the horrified silence. "This same course of ingestion followed by rapid, violent allergic reaction occurred whenever a Terran animal consumed Thorbian flora or fauna. Results were immediately fatal and even having anti-histamines ready to administer was no guarantor of recovery or survival."

I brought up a slide showing two slightly different globular structures. "This Thorbian protein and the corresponding Terran protein both perform the same function in their respective organisms. Note the slightly different surface conformation; as far as we can determine, our bodies should be able to digest the Thorbian version with minimal issues. The problem seems to be our Terran immune systems treating every Thorbian protein as a major allergen. Since eating Thorbian vegetation floods a body with Thorbian proteins, our Terran immune systems will go into rapid onset anaphylaxis with every meal." In the dimness, I could see eyes that had been riveted on the screen suddenly focused on me, hoping for a solution.

"To be thorough, we placed containers of Terran grass beyond The Glass for a period of two weeks." I brought up a slide of five pots of grass with a robot carefully tending them. "The grass was harvested and brought back inside and randomly split into two treatments. Treatment A was washed repeatedly, then fed to lab rabbits. They exhibited no adverse side effects. Treatment B was not washed and fed directly to the rabbits. They showed a less severe allergic reaction, suggesting that Thorbian pollen..." I paused and flipped to a slide showing the pollen grains in question, "...caused the reaction.

"We attempted immunotherapy using the most common species of Thorbian grass in the area. Some lab animals could safely eat it if we washed all dirt and pollen from it before feeding. If we attempted to feed any of them unwashed grass, or another species of plant, we triggered another attack of anaphylaxis. At this time, we consider it too dangerous to attempt similar inges-tion experiments on humans, though a skin test was performed on a volunteer from our lab under the medical oversight of Dr. Arun Ramakrishnan." I flicked a rumpled Arun a quick glance.

The next image showed a human arm from hand to mid-bicep, swollen and mottled red and white with fingers plumped up like sausages. "Typically, a scratch test takes about twenty minutes for allergic symptoms to occur. The above occurred in two minutes. Antihistamines were administered, but it still took over twenty-four hours for the symptoms to completely disappear."

I exited my presentation and used my tablet to raise the light levels in the room. "Any questions?"

Captain Eugenia Richardson let the cacophony go on for a bit, then slammed her palm on the table. "Enough!" The room fell instantly silent, as if it were a lab of rowdy undergraduates being scolded by their professor. "Better." She turned her attention to me. "Thank you for presenting your findings today, Dr. Morgan. This presents a significant challenge, one that I've not heard of in any other colonization effort. What safety equipment would a person require to go outside The Glass?"

"A minimum of a Biosafety Level Four suit. Full body enclosure, air tanks, everything, with a full decontamination upon returning inside."

From my left, a nasally tenor voice broke in. "Are you sure it was anaphylaxis? Maybe you just fed them something poisonous and killed your little bunnies that way."

The voice belonged to Dr. Mark Dan, one of the physicists. Lovely. He was an ass both as a person and colleague, constantly doubting the professional knowledge of others.

"Dr. Dan." I turned to address him, striving to keep the venom out of my voice. "Computer modeling based on the underlying structure did not indicate that any of the plants used in the experiments produced toxins. The plants in question were observed to be consumed by multiple species of Thorbian herbivores without incident. To the best of our knowledge, the plants used in this experiment were not producing any toxins that would interfere or amplify the allergic response recorded." I gave him a small, tight smile that didn't reach my eyes.

"If you are truly concerned about the rigor with which I conduct my experiments, I invite you to observe the next post-mortem of an experimental animal so that you may satisfy for yourself that the technique and diagnostics were performed to accepted professional standards." He would never do it, but the idea of Dr. Dan getting his hands dirty with what he had loudly proclaimed to be a "soft" science was entertaining.

"Dr. Morgan." It was the captain, and there was a slight edge to her voice. "Thank you for your efforts and expertise. This is a serious problem, and one for which we must find a solution. We still have twenty-five hundred colonists in stasis aboard the *Virginia Dare* and we cannot keep them up there indefinitely."

"It would seem that they are safer than we," interjected Dr. Dan darkly.

Dr. Ramakrishnan spoke up. "Mortality rates, even in these conditions, wouldn't lead to the extermination of the colony, Dr. Dan. We will, however, have to adjust our expected population growth rates drastically downwards."

Dr. Dan would not be quelled. "If this planet is so dangerous, why can't we just leave? Put everyone back into stasis, scrub the expedition, and move on. We needn't confine ourselves to this world."

"That is not an acceptable answer, Doctor," the captain said, her voice cold. "We have all spent two years in stasis to get here, and the rest of the colonists remain that way. We are rapidly approaching the point where the risks of long-term stasis begin to rise dramatically. The next potential site is nine months away. Even if there are no issues with that biosphere, the likelihood of stasis-induced brain damage, immune system disorders, and death exceed acceptable levels. We could lose twenty percent of the colonists just trying to get there and still fail."

She looked around the room, as if daring anyone to challenge her logic. "The main colony fleet will be here in three standard years. That's twenty thousand colonists who have given up everything to reach a new home. We cannot, and will not, abandon them to their fate. Lives depend on us finding solutions. This is why you were all selected for the advance team and why you are being paid so well. We don't have time for panic or despair. I need you all to put your shoulders to the wheel and bring me solutions.

"You have sixty local days. Let's get to work." Without another word, she rose and left the conference room.

A few other researchers came to me with questions, but I felt like a deflated balloon. The entire colonization effort was in jeopardy. I didn't know how to overcome this, or if it would even be possible.

❖   ❖   ❖

Later that night, Alan and I lounged on the floor of my quarters eating a meal he brought from the cafeteria. I didn't feel like having my mealtime interrupted by more questions, so we retreated to my room to eat.

"So, we're allergic to an entire planet?" Alan asked.

"We are. Even if we don't eat anything, the pollen and other biological matter in the air will cause us to react and die. Unless we figure out a solution, our only options are to remain confined to artificial habitats or scrub the whole expedition. I don't think staying inside The Glass is viable, either, not in the long run. We'll be packed in here like cattle... or like a prison. Socially, the colony would be miserable."

"Tell you what, babe, I found an awesomely cheesy sci-fi flick in one of the entertainment files. Let's relax tonight—you be can be busy starting tomorrow."

I nodded, relieved to think about something else for a while.

The next morning, Dr. Ramakrishnan waited for me at my lab. He said, "You certainly got your point across yesterday. I noticed more than a few people going green during parts of it."

"Hey, you said 'make sure everyone in the audience understood the scope of the issue we were facing on Thorbia,'" I replied.

"Based on the reaction to Dr. Dan's proposal, you may have succeeded too well. Before there was anticipation, now there is fear. We must find a way to turn that into hope."

"I'm a bit short on any long-term solutions—much less hope— this morning." I was truly at a loss for options.

Arun smiled. "Today we shall let our ideas rattle around in each other's brains and maybe we will find a solution that way."

As we walked through the lab toward my office, I gestured for Tara to join us. Soon we were all settled, or slouched in Arun's case, while I scribbled notes on the smartboard. With the help of caffeine and camaraderie the ideas began to flow.

Tara, who up until now had been silent, suddenly perked up and asked, "If the problem is the body's immune response, then why can't we genetically alter it to ignore the alien proteins?"

For the first time ever, I saw Arun sit up straight in his chair. He said, "We'd have to target the genes of the major histocompatibility complex classes of molecules so that they treat the Thorbian proteins like Terrestrial proteins and not trigger anaphylaxis."

"Major histocompatibility complex classes? That makes sense, those are the molecules responsible for triggering immune responses." I wracked my memory. "Aren't those started in the thymus?"

"Yes, you will be altering the expression of genes in a critical part of the immune response. Worse, we'll leave ourselves vulnerable to whatever Thorbian diseases exist," Arun warned.

Finally, Tara spoke. "All of us in this office know that there are always trade-offs in a biological setting. Frankly, the possibility of death out there is better than the certainty of death stuck in here."

"Can we do genetic modifications of this type and on this scale? And more importantly, can you make these changes inheritable?" Arun asked.

"Yes," I replied. "We knew that we would need to genetically modify our food crops and animals as well as manufacture vaccines for a growing colony. We will have to refine some of the production processes when it's time to start human trials. The hard part will be transplanting parts of the Thorbian genome into the Terrestrial one to make sure both conformations are produced."

Arun looked me in the eye. "You know there will be unintended, long-term consequences from this. Any complications, any failures, and we will be blamed. We can model this all we want, but you and I both know that reality may not match what the computer says. Are you prepared for what might come next?"

"We don't have a choice," I responded grimly. On Earth, we used CRISPR-based techniques to do targeted edits of the human genome to eliminate certain genetic diseases. But that was as far as it ever went. Public opinion was still strongly against editing the human genome as easily as a potato's. Those who didn't face the hard choices presented to us would judge us with the clarity of hindsight.

"Yes, let's start working on this."

I gave a slow nod. "Dr. Ramakrishnan, would you start working on setting up the clinical trials? Dr. Hauff, you and I need to start finding the Thorbian analogue to the major histocompatibility complexes. We need to set out more small animal traps—we're going to want multiple individuals from multiple species."

This would be the hardest part. We would need to induce an immune response in multiple Thorbian animals to find out how it presented and what the source was. This would require blood transfusions between Thorbian species and observing the results. There would be no progress without sacrifice.

I woke to my shoulder being shaken. "Mmmph?"

"Molly, wake up. You can't sleep here." Alan came into view as my eyes focused, concern on his face.

"No, babe, I'm fine." I mumbled, trying to reassure him. "I need to check on a few things in the lab, then I'll go bed."

"Tara tells me you've been out for over an hour. Come on, let's go to your quarters. I brought dinner for us. Eat something and then you can get some uninterrupted sleep."

In the doorway, I saw Tara looking in at us with a satisfied grin on her face. *Traitor.*

It had been another long day in the lab. The captain's deadline was fast approaching and it was an all-hands effort. Everyone was working in shifts, even with Arun sending over help as he could spare it. I had intended just to put my head down for a catnap before getting back to work. But Alan's offer of food and an actual bed was too tempting.

"Go sleep, Molly," Tara gently commanded. "The lab can manage itself until you get back. I promise that if we get a breakthrough, then I will let you know immediately. Besides, you need to look alert when you present our idea to the captain in two days."

"What?" I responded, my brain still waking up.

"The proposal for the captain to determine the future course of the colony is in two days," Tara reminded me. "You need to put together the presentation. I'll edit some footage tonight and you can finish it in the morning."

With a warm hand on my lower back, Alan gently nudged me up from my chair and herded me toward the door, with Tara smiling all the while. She knew I couldn't resist Alan, who would pick me up and carry me if necessary.

When we reached my door, I caught a whiff of the food Alan brought over and my mouth watered. I couldn't remember the last time I had eaten. The mass-produced cafeteria food smelled amazing right now.

After a few minutes of stuffing my face, I asked Alan how things were going outside the lab. "The closer the deadline gets, the more calls my team gets. People are scared and tempers are flaring. We've had fights, domestics, you name it."

My hand reached to cup his face across the table. "I'm sorry, babe. I've been so caught up in my own research I hadn't realized things were so rough out there."

Alan shrugged. "We're all in this together. This sort of situation is why my team is here. I will say, though, some of these scientists are surprisingly spry when they're trying to beat on each other." He grinned. "What have you been working on? You've barely left your lab for the past month. Have you found something?"

"Arun, Tara, and I are trying to find the source of the immune response in Thorbian animals. Then we'll try incorporating that into our human genes so that our immune systems don't freak out when they encounter Thorbian proteins. It's a long shot, but we think we can make it into a long-term self-sustaining solution."

"So, you're going to alter our DNA? That's wild. That sounds like the plot of a horror movie. Should I warn my team of an outbreak of zombies or mutants?"

I chuckled. I'd needed the laugh. "In all seriousness, the possibility of something horrific happening is why we're running closely monitored clinical trials on lab animals, including monkeys. Then we can attempt to do this on humans."

I yawned again. Alan tugged me toward the bed. Tomorrow would be a busy day as we assembled the team's presentation.

The room was packed beyond maximum capacity and the air-conditioning was struggling to compensate. A wide range of solutions had been presented, from burning everything upwind of the landing site and seeding the land with Earth vegetation in a sort of slash-and-burn terraforming, to tunneling into the Thorbian dirt for future expansions.

Then it was Dr. Dan's turn. Smartly dressed, he gave a smooth presentation filled with slick animations. He waxed rhapsodic about how the entire colony could band together, and if we all sacrificed some part of our personal research efforts, we could help build a larger enclosure. His plan used orbital scans to find appropriate mineral deposits and set up remotely controlled mines and small-scale refineries to produce more materials and expand our current habitation.

The room seemed to gain a sense of hope at his plan, but I wondered just how much everyone would be expected to sacrifice to Dr. Dan's control to see his plan come to fruition. When the opportunity arose, I asked, "Dr. Dan, exactly how would the use and allocation of labor and robots from each department be determined?"

"In the event that my proposal is accepted, I will draw up a roster that will fairly distribute the load across all the departments and personnel based on the needs of this project."

Pompous jerk.

"You'll draw up the roster?" asked Dr. Ramakrishnan, raising an eyebrow. "Surely such matters are the purview of the captain as the duly appointed representative of the Starpherra Corp?"

"Ah yes, you are correct Dr. Ramakrishnan." Dr. Dan's cheeks reddened slightly at his loss of face. "I should have said that the captain and I would work together to determine a suitable roster and logistical plan."

I wondered how easy presumptively ruling with an iron fist came to Dr. Dan and how much of my distrust was due to my personal dislike of him.

Shortly thereafter, it was my turn. I had tried to dress professionally, and had even put on makeup, but I still felt nervous as I began. "Captain Richardson and my fellow members of the Thorbian expedition, the plan that my department, along with the help of Dr. Ramakrishnan, has developed looks at how we can alter our immune system's response and thereby treat a day outside on Thorbia like we would treat a day outside on Earth."

My fingers trembled as I played the first of the videos that demonstrated how our immune system essentially worked the same as the Thorbian one and explained how, through careful experimentation and computer modeling, we determined the source and pathways of the Thorbian immune responses. I detailed our plans for the comprehensive modeling on which our genetic modifications would be based. Finally, I reached the last slide.

"I propose that instead of expanding this gilded cage we currently live in, we escape it. We must use our current biotech capabilities to change ourselves and our children so that we may freely seek the Thorbian horizon and experience this planet directly for ourselves."

Dead silence met my proposal for two very long heartbeats, then the room erupted into shouting.

I glanced at Captain Richardson, but seeing that she was too deep in contemplation to restore order, I turned up the volume on the mic and announced, "I cannot respond to your questions and comments if they are not presented in a coherent and professional manner."

The throng quieted and I started answering the questions that came flying toward me. Some were technical, while others reflected the individual's own worries and beliefs.

Soon enough, Dr. Dan spoke up. "What if this turns us into inhuman mutants over generations?"

"Then we will have bought ourselves time to find a more permanent solution."

"Not if it also turns us into gibbering morons," he shot back. Behind my back, I clenched my fists. I couldn't lose my temper.

"Dr. Dan, none of us in this expedition are scientific slackers. I and Dr. Ramakrishnan will perform the required tests and clinical trials so that whatever solution we produce will be to the standards that Terran regulatory agencies require of any gene therapy. But I acknowledge that every human has the potential to respond differently to any medical treatment that exists." I tried to rein in my temper and felt my shoulder muscles ache from the tension. "Unlike you, Dr. Dan, I have dealt with the potential unpredictability of biological systems and accept that sometimes things cannot be modeled perfectly within a living and changing system, unlike when dealing in a purely academic laboratory situation." Perhaps that had been a bit too pointed.

Captain Richardson spoke up. "I believe this discussion has gone on long enough. Send me your complete proposals by midnight tonight, along with a cost and material breakdown. I will review the files and let each of you know my decision in a week." Captain Richardson stood up and turned to leave the conference room, cutting through the crowds like a battleship underway.

"A week? What are we supposed to do until you make up your mind?" Dr. Dan demanded.

Captain Richardson turned and said coldly "I expect you to continue doing your jobs."

In the silence that followed her exit, I gathered up my tablet and hurried for one of the side exits. I had a lab with numerous experiments to get back to and another three pages to add to my proposal.

✧   ✧   ✧

True to her word, in a week's time Captain Richardson called me into her office. "Have a seat Dr. Morgan. You look exhausted." Her voice was warmer in private. Her office was hung with photos of Earth and Thorbia, and the back wall had two large windows overlooking the landing area and the Thorbian wilderness. On her desk a pair of binoculars were tucked next to the computer.

"Thank you, Captain Richardson," I replied as I settled back into a surprisingly comfortable chair. Gesturing to her binoculars I asked, "Do you ever see anything interesting out there?"

"Oh, lots of things," she smiled, "One day it would be nice to experience them as well. Which is why I called you in here. I have made my decision. We will pursue your solution first. It is the only one that, if successful, will allow us to integrate the colony into the planet."

I sat in stunned silence for a few seconds then shot to my feet. "Yes, thank you Captain Richardson," I reached across her desk to shake her hand, bouncing in my excitement.

Disengaging her hand from my enthusiastic shaking, Captain Richardson bade me to sit back down. I did, this time on the edge of the seat cushion. "Now," she started. "How are you coming along with your research?"

"We've begun working on the method of inserting the Thorbian DNA into the rat genome. We're using the existing CRISPR methodologies that worked well back on Earth. Unfortunately, they've never been tested with Thorbian DNA."

"And has your...what did you call it? CRISPR? Has it worked?"

"Eventually." I gave her a slight smile. "We had to add a bit of Earth DNA to the ends of the Thorbian pieces so the enzymes would have something to attach to. Otherwise the presence of the extra genetic bases caused it to ignore the genes we wanted it to splice in."

"Excellent. But did it *work*?"

"We don't know yet. We didn't want to get too far into the clinical trials before we knew what your decision would be." I was suddenly reminded of Arun's concerns. "Captain, what we're trying—this genetic modification—it will be a lot for some people to cope with. There was still a strong anti-genetic-engineering crowd on Earth when we left. Even though we're all employees of a multinational pharmaceutical company, there are still people who are reluctant to have this done to them."

"I know, Dr. Morgan, I'm one of those 'reluctant people,' but at this time I don't see how we have a choice. Right now, we live at the mercy of our equipment, and it is only a matter of time before something fails." Her eyes flicked to the photos on the wall. "And I have begun to wonder if it is our choices that define our humanity, and not just our DNA. Are we truly human if we come this far only to choose to live in a gilded cage of our own making?"

"But what about the people who refuse this treatment?" I asked.

"As captain of this ship and duly appointed representative of the Starpherra Corp., I am empowered to back my decisions up by force, if I deem it necessary." Her voice was back to its typical no-nonsense tone. "I will make the announcement of my decision within the hour. I expect you to send me weekly reports of your progress. Good luck, Dr. Morgan," she said, dismissing me.

"It looks like this round of trials on the rats have a higher success rate at least, but they're still sickening after a while. They're eating the Thorbian vegetation without going into anaphylaxis, but then they get run-down and bloated, then listless and finally die." Tara ran her hands through her hair, a sure sign she was frustrated at a lack of answers after six weeks of failures.

Dr. Ramakrishnan poked gently at the organs of a dead rat as he examined it. "These all look paler than what they should be. And while the animals are bloated, it seems to be water weight. Look," he said and indicated the diaphragm muscles. "These look almost atrophied."

He leaned back from the postmortem on the lab bench. "Have you done an analysis of their feces to determine digestion of the Thorbian proteins?"

"What are you thinking, Arun?" I asked, as I sat upright on the opposite side of the exam table.

"We hypothesized that despite the different conformations, our Terran gut biota would be able to digest the Thorbian proteins. What if they can't? Or at least, what if they can't digest them completely? Maybe the rats don't die of anaphylaxis; instead they die of malnutrition."

"Tara, have we examined the fecal composition?" I asked.

"No, we haven't. Not yet at least. But I'll get someone on that right now," she said as she rushed off.

After a few hours, the results came back: Only about fifty percent of what the rats were ingesting could be broken down and absorbed by their bodies. The Thorbian food filled their stomachs, and so they thought themselves satiated, but their bodies couldn't digest enough of what they ate.

To populate our Thorbian modified rats' guts with the right digestive bacteria, we collected fecal samples from Thorbian animals with similar diets. We then put small amounts into capsules and fed them to the rats.

"Making poop pills for rats isn't the weirdest thing I've done for science," Tara commented. "But it certainly ranks in the top ten."

"Let's hope these fecal transplants work," I said. "I want to move on to the next set of trials."

The autopsy of a subsequent rat a few weeks later revealed a creature in much better health, though with a slightly enlarged cecum. Concerned about possible infections, we ran more tests.

"Congratulations Drs. Morgan and Hauff, you have now given the appendix a reason to exist," announced Arun. "It seems that the Thorbian gut fauna moved into the rat's version of the appendix."

After a round of celebratory elbow bumps, we started setting up multiple trials on larger animals. We needed a large enough sample size to draw meaningful conclusions based on the data.

After a few months of successful tests on rabbits, we moved up the chain to monkeys. Monkeys are much more like humans, so the data could be more easily extrapolated to us. They would clearly show if the Thorbian digestive bacteria still showed up in the appendix. Unfortunately, monkeys were a very limited resource, so any failures had a greater impact.

"We're getting close, babe, I can feel it," I told Alan while snuggled in his arms. "The monkey trials had good results. There were a few odd side effects in three of the macaques, but everything was still within acceptable statistical limits." I grinned up at him.

"I'm on the security squad, not the math squad, honey. I'm guessing based on your grin and body language what you just told me is good news?" He peered down at me quizzically.

"Yes, what I said is very good news. It means we're closer to a solution that will allow us to survive outside of the ship.

Plus, all this data we're gathering now can help us with future experiments," I explained.

"So, what happens when you get ready for human trials? How do you determine who gets the first injection?" he asked.

"We have run genetic compatibility and immune-response trials against tissue samples from everyone in the crew. We're going to randomly select one of the most promising candidates to be the first guinea pig," I said.

"That sounds like the worst lottery ever," Alan chuckled.

Ten weeks later, it was time to choose from the hundred best candidates. Tara started the randomized algorithm she had designed, and, in a few seconds, we would have our first experimental human. I hoped whoever was chosen would understand the importance of this experiment and cooperate willingly.

Dr. Hauff read out the name and I looked around expectantly, excited to see who was chosen. But everyone was looking at me. It was then that I realized the name that was read had been my own. At least willing cooperation wouldn't be an issue, I thought darkly.

Now to break the news to Alan.

Alan had not been happy at my winning this particular lottery. His protective nature railed against the thought of the potential dangers but held firm. If I tried to weasel out of it, everyone would think I had no faith in my plan. If I were willing to risk the future of the colony, then I must be willing to risk myself.

"Molly, I . . ." his voice was thick with emotion. "Just be careful, please. I can't imagine my future without you."

Three weeks later, it was finally time for the human test to begin. I showed up at the medical wing to no small amount of curiosity. Everyone knew why I was here, and everyone wanted to know what would happen. Alan managed to get a few days off and was adamant that he would be staying by my side throughout the process.

"So, Arun, this is the wrong time to tell you that I hate shots, right?" I joked. A mixture of nerves and excitement were making me talkative.

The treatment I was getting had been tailored to my genetics to maximize effectiveness and minimize any possibility of rejection. All that remained was for Dr. Ramakrishnan to administer the series of shots, give me a fecal transplant pill, and keep me under observation for forty-eight hours. After that would come the CAT scans, MRIs, EKGs, and any other test the good doctor thought he could get away with.

I swelled up at each of the injection sites, but topical antihistamines and an anesthetic cream kept it from being more than just a bit of discomfort. I swallowed the poop pills quickly while trying not to think about what they were. Over the next two days I felt a little something like heartburn on and off and a twinge or two in my lower abdomen, but mainly I slept. Or at least I tried to, despite the steady stream of medical personnel coming in and out of the room to collect data and samples.

Every time I woke, Alan was there, waiting patiently. We would talk while I was awake, making plans for the future.

Eventually Dr. Ramakrishnan came in with a wheelchair. He and Alan shifted me from the bed to the chair—despite my repeated protests that my legs worked perfectly well. Arun then measured me in every conceivable way. I soon felt I had more wires attached to me than most computer circuits. I wondered if my racing heart was skewing the tests.

"Molly, the MRI is showing a slightly enlarged appendix like what we saw in the monkeys. Are you in any pain? Is there any tenderness?" Arun asked.

"None, Doc. Imagine coming all the way out here to finally find a use for the appendix!" I laughed, hoping I was done with the medical tests.

"Otherwise, Molly, you are in good health and seem to be showing no ill effects. It's time for the final test. I have the medical team standing by."

"Okay, Arun," I looked over at Alan. "Babe, will you wheel me over to the exit hatch and help me suit up?"

"Yeah, and I'll be double-checking your seals, straps, and air tank. I'm not trusting anyone else to do it." There was no arguing with Alan on this.

Even with another person there to help adjust the straps and run the air and radio checks, putting on a biosafety suit still takes time. After a final, lingering kiss, Alan sealed the last zipper shut. I moved to join the medical crew at the entrance to the room that housed the emergency exit.

I keyed my throat mic, "Guinea Pig One is ready to proceed, over. Is Team Safety Net ready to proceed? Over." I got a chorus of affirmatives from the four people on the medical team.

The door cycled opened and we stepped through, then waited for it to close and seal behind us. The emergency door opened, and one by one we stepped through and out into the Thorbian sunshine.

We only walked a few meters away from the habitat—still close enough if they had to drag me back inside for medical treatment.

"Guinea Pig One in position, over," I said using the radio.

"We see you Guinea Pig One. Safety Net at the ready? Over," I heard Captain Richardson's voice respond.

"Safety Net ready, Captain Richardson. Over."

"Proceed with the experiment. Over." When I heard Captain Richardson's command, I slipped my hands out of the oversized gloves of the suit and removed the ventilator from my mouth.

Fumbling with the external zipper of my suit, a small part of my brain marveled at how calm I felt.

Showtime.

I began opening my suit, exposing myself to the Thorbian atmosphere.

One breath, two breaths. I felt no closing of my throat, no racing of my heart. I continued to breathe for a few more minutes under the watchful eyes of the med team.

I cued my mic. "This is Guinea Pig One. I feel no ill effects. Over."

In my earbud, I heard the cheers from the Safety Net team. This was it, we had broken free of the gilded cage. Now we could chase that far distant horizon—we would not just live on the planet, we could become part of it as it had become part of us.

Biological and Medical Challenges of the Transition to *Homo Stellaris*

Nikhil Rao, MD

Nikhil Rao, MD, MSc (AKA Vindaloo Diesel) is a psychiatrist who works with critically and chronically ill kids, making him the only physician in the pediatric ICU who can dress up like Batman and claim it's "for the children." He also develops new psychiatric interventions for children with serious medical illnesses. Prior to medicine, he studied the evolution of monkeys, which are a lot like children, albeit hairier, followed by some time in the fitness professional community during which he proudly earned the dubious badge of "Strong for an Indian." He's unsure where his creative career is heading, but it probably involves a juvenile sense of humor, a sprinkle of profanity, and entirely too much math.

INTRODUCTION

In the next few hundred years, it seems likely that at least some of us will reach for the stars. Whether by desperation, aspiration, or simple curiosity, humanity has always reached for distant lands. Challenges and opportunities are manifold, and while often discussed at technological and social levels, the biological and psychological considerations are no less interesting or important.

Unlike physics, and to a lesser degree chemistry, the science of biology and profession of medicine are only a few hundred

259

years old in any substantive sense. Given medicine's youth and the rapidity of change, even now, the coming centuries will doubtlessly generate technologies that will seem miraculous. Yet, certain biological realities and psychological imperatives are unlikely to yield to mere technology.

Here we explore medical aspects of colonist selection, life aboard a slow-traveling colony ship, and the colonization process. We will also consider potential technological targets for intervention.

THOSE WHO LEAVE

The possibility of faster-than-light travel cannot be discounted, although presently no proposed technology or translatable mechanism exists. Quantum entanglement offers a realistic mechanism for instantaneous communication, but transmission of matter through similar means remains speculative to say the least. Thus, we will proceed under the assumption that while technological advances like solar sails or nuclear rockets may increase the speed of interstellar travel, it will remain frustratingly slow, with travel across light-years measured in decades or centuries at best. For similar reasons, it is best assumed that colonization missions are one-way trips, with resupply or rescue effectively impossible. Chosen individuals must be thoroughly screened, with a range of attributes which may not be immediately apparent (see Robert E. Hampson's story, "Those Left Behind").

Technologies such as CRISPR/Cas9 already allow gene editing, with human trials of direct modification of the genome already under way for treating such diseases as cystic fibrosis. But there is a major difference between single-gene engineering and modifying the expression of complex traits such as intelligence, physical strength, or longevity. In the latter cases, hundreds of genes contribute, with gene-environment interactions and differential effects based on time and age of exposure. For example, over one hundred genes contribute to the expression of intelligence, and over a thousand to immune functioning. Consider also that epigenetics—changes in how strongly our genes express themselves in response to environment and experience—alters the simple inheritance of traits. A single episode of physical or emotional stress in infancy often leads to reduced intelligence and stature, emotional lability, and compromised immunity that may prove lifelong. However, in adolescence/adulthood, periodic episodes of physical stress (when in an appropriate recovery framework) promote improved functioning across those same attributes.

Ultimately, most positive traits in humans are emergent functions of genes, environment, our interactions, and time. While gene manipulation and nanotechnology may modify these processes, potentially eliminating negative traits, they will likely not change the fact that human traits are ultimately distributed along a series of bell curves, even as science shifts the shape of those curves.

Physical Robustness

Obviously, anyone picked for such an arduous, resource-intensive, high-risk, long-term task must be physically healthy. The specifics of what that entails bear consideration. General "fitness," longevity, and low risk for chronic debilitating disease go without saying, all of which, even now, are either directly measurable or predictable. On the other hand, health issues so mild they may be safely ignored on our home planet may turn deadly as we reach for the stars. The question remains as to what is "fit" and what is "ideal," as different environments determine the optimal parameters.

Immune Function

Consider the immune and inflammatory system. Many people have mild seasonal allergies or a tendency toward contact rashes which, while aggravating, don't severely impact daily function or overall wellness. Allergic response is not intrinsic to these otherwise benign substances in the environment; rather, the body aggressively inflames and destroys its own tissues in response to a perceived threat from "foreign" substances such as proteins on pollen, bacteria, or viruses. Extraterrestrial organic chemistry and potential life forms, being literally alien, will likely register as even more "foreign," apt to provoke more severe, even deadly reactions.

Mild immunodeficiency is surprisingly common. Many people who "get sick every year" with the same organism are clinically deficient in specific antibody or immune-cell function. Often, it's the antibodies lining tissue exposed to the environment (mouth, eyes, gut) that serve as a first line of defense, which broadcast to other immune cells a virtual call-to-action. Considering the prevalence of immune deficiencies, the threat of novel extraterrestrial pathogens and the hasty evolution of Earthly pathogen stowaways, these normally inconsequential pathologies could turn deadly.

Technological solutions, whether through genetic engineering or artificial immune systems, would have to grapple with the same issues of immune under- and overactivity.

Physical Characteristics

Skeletal robustness and muscle fiber composition essentially determine an individual's optimal size. Muscle characteristics then dictate endurance capacity and relative raw strength. All of these factors impact colonization suitability yet are likely difficult to manipulate with significance through technological solutions. In addition, these same features dictate nutritional needs and metabolic waste.

Since the earliest days of space exploration, ship volume and mass with respect to the available biological load have been key constraints on the distance and duration of the journey. Technology will impact cost and efficiency curves of these factors, but they will likely remain significant variables.

Biological waste production aboard ships includes CO_2 as well as liquid and solid excreta. The amount produced by an individual directly relates to caloric intake and energy expenditure. The more mass someone has, the more they will expend as a simple function of having more metabolically active tissue. Higher-mass individuals are often Type-II (anaerobic) muscle fiber dominant, resulting in higher rates of resting and active metabolic rates. Thus, nutrition, oxygen production, and waste turnover are heavily impacted by mass, meaning larger, more musculoskeletally robust individuals will be difficult to provide for both per unit of mass and per individual.

Finally, target planets are likely to vary considerably in climate and gravity and early colonists will likely have to deal with these elements (or adapt to them) to some degree, regardless of what habitats and shelter they bring with them. For example: Heat production is a simple function of metabolic rate. Heat dissipation and retention, however, is a function of surface area to volume ratio; density retains heat. A classic example of adaptation to environmental extremes is the short-stature, thick-torso Inuit who hail from icy terrains, versus the long and lanky Masai of the hot, dry savannah.

The gravity of a colony world will likewise have physical implications. We know from prior space missions that microgravity of just a few months duration can dramatically affect biology, sometimes yielding beneficial effects and sometimes, not so much. Sustained conditions in higher gravity will have several anticipated harmful effects. The heart does its work against gravity, either directly (as in cerebral flow) or indirectly through the lower circulation. The venous system must be able to maintain tension and competence against the blood contained within. When this process goes awry, we develop varicose veins, stasis ulcers, and clots. Connective tissue is also affected, with the strain of gravity amplifying the degenerative weathering of arthritis and tendinosis. Taller individuals will struggle, while more gracile individuals of any height will be ill-equipped for the extra apparent weight. It is difficult to imagine the optimal human build for a very warm, high-gravity world. Perhaps such worlds would be poor targets for colonization.

Lighter gravity imposes fewer constraints on colonists' builds and increases potential physical performance discrepancy. Relative gait or strength differences in differently sized individuals would become greater absolute differences. And, with enough time, colonists inhabiting lower-gravity worlds will assume less dense, less resilient bone tissue, which while perfectly fine on their planet, may constrain or prevent travel to higher-gravity planets; and, if artificial gravity continues to evade us, circumstances might limit them to slower travels due to lower tolerance for spacecraft-provided thrust.

SHIPBOARD LIFE

Regrettably, based on current knowledge, colony ships are unlikely to achieve velocities at any appreciable fraction of light speed. Given a maximum of $0.1c$—achievable by solar sails and fusion propulsion—a trip to Alpha Centauri (with no proven habitable planet, but likelihood of such estimated at eighty-five percent) would take an estimated 130 to 150 years, well beyond the current human lifespan.

Lifespan

Molecular biology and medicine have taunted us with numerous longevity breakthroughs, none of which have survived from the short-lived mouse model to long-lived humans. While we broadly understand determinants of aging and chronic disease, operationalized interventions with meaningful results remain elusive. Some significant factors implicate telomeres, endcaps of chromosomes, that shorten with each cell division, bringing us ever closer to the day our bodies are unable to replace senescent cells.

Telomerase, the enzyme that facilitates telomere lengthening, is useless to prevent nerve and muscle-cell aging. Those cells, with few exceptions, are the ones with which we were born. In such cases, cellular repair mechanisms simply cannot outcompete the inevitable damage, whether by mechanical means, toxins, or exposure to space hazards. Spontaneous gene mutations also wreak havoc. One mutation may allow a single undetected cancerous cell to become an incurable tumor or render a crucial cell lineage ineffective. These processes are targets for longevity technology advancements through a myriad of current and potential modalities such as direct intervention (gene editing) for more robust cell protection mechanisms, monoclonal antibodies ridding the body of anything cancerous or otherwise mutated, or through nanomedicine working to manage the toxic byproducts. The latter intervention is particularly important when one considers that most dementias are related to buildup of substances in the brain that the body simply cannot eliminate. Technological process will certainly lead to longer, healthier lives. However, achieving biological immortality, or individual lifespans compatible with a multicentury journey aimed at transporting the original crew to terraform and colonize a new world, is not yet clearly within our grasp.

Microgravity and radiation health concerns

Since the dawn of human spaceflight, we've seen how microgravity and radiation correlate with a range of health concerns, with rapid deconditioning occurring within days to weeks. Intervals of relative weightlessness, as seen with Mir and the International Space Station, pose other problems. Notable changes associated with microgravity include rapid bone-density loss, accelerated atherosclerosis, and thus greater heart attack/stroke risk, kidney impairment, and intracranial pressure changes, which may accelerate age-related brain and ocular damage. Many such factors relate to losing gravity's prominence in organizing passive directional movement of various body fluids and, unfortunately, on-board strength and endurance training will not likely prove entirely effective at mitigating these risks. We will almost certainly require some mode of gravity replacement, even something as simple as rotational or thrust-induced methods.

Shielding will be paramount in any future missions. The primary current limitation on simple physical shielding is the cost of placing mass in orbit. We must solve this technical limitation before constructing colony ships. Other solutions, such as using hollowed-out asteroids as shields or as hulls, or developing artificial magnetic shields are worth considering. We understand the processes to execute the latter, and improvements in technological accuracy, transmission, and power production will likely make this possible.

Biologically speaking, the radiation-damage problem is complex and daunting to solve through genetic or proteomic tinkering. Cancer is just one of many bleak radiation risks. Others include increased neuronal senescence, affecting neurodegenerative disorders and dementia (remember, neurons generally are not replaceable). Senescence can occur in progenitor cell lines as well posing a threat to connective tissue, bone, and immune-cell maintenance. Fortunately, the body is rife with DNA repair mechanisms and proteins designed to soak up radiation and other damage (sometimes called heat shock proteins). Conveniently enough, these are more robust or more numerous in other organisms. The water bear, a tiny arthropod form the phylum *Tardigrada*, is famously resistant to radiation (and many other environmental hazards). Scientists have already proven that its protective proteins and DNA repair mechanisms can protect human DNA (in petri dishes) from the ravages of radiation.

To sleep, perchance to dream

Cryonic storage, along with other forms of suspended animation, has captured the imaginations of science fiction authors and the wallets of wealthy, death-fearing individuals since scientific inquiry began to burgeon in the 1950s. Unfortunately, true cryonics has proven moribund. Simply put, freezing the body turns water to ice—crystalline and sharp—which does unkind things to cellular structures. Moreover, beyond the simplest of simple multicellular organisms—namely, the ubiquitous and nearly indestructible water bear—there has never been an inkling of success at bringing an animal to a complete molecular halt and then simply restarting it. On the other hand, slowing metabolism is successfully used clinically already, albeit in the rather limited and coarse fashion of using ice baths to induce hypothermia. This has led to an increase in viability and time to intervention in situations such as sepsis or cardiac arrest in which the body's cells are rapidly dying. These periods are measured in minutes to hours, not decades to centuries; however, imagine if they *could* be extended to such long periods (see Kevin J. Anderson's story "Time Flies").

This leads to the tantalizing and perhaps more plausible possibility of hibernation, a strategy used successfully by several vertebrates—whether lungfish surviving the dry, hot summers or bears dealing with nutrient-poor winters in arctic and subarctic climes (see Les Johnson's story "Nanny"). Hibernation is a logical target for research for a number of reasons:

1. Mammals already do it, meaning there is an inherent biological feasibility either through gene editing, protein/hormone administration, or both.

2. Hibernation has clear endogenous (hormone and blood protein) triggers for induction and exit.

3. Hibernation dramatically slows metabolic processes including energy requirements and cellular turnover, which in turn delays metabolic waste accumulation and cell damage.

Moreover, studies of animals that facultatively hibernate (meaning that they do if the winter is cold or food scarce) reveal that hibernation correlates with longer lifespan in addition to reduced caloric intake. Bear hibernation is particularly fascinating, as it occurs without loss of lean mass or brain density, in addition to a total lack of production of urine or feces, in a fairly long-lived animal.

Hibernation could reduce caloric needs by up to ninety percent based on animal models and produce up to a ninety-percent lengthening of lifespan at the theoretical high end. Simply stated, a month of lifespan is earned for every year in hibernation. Ten individuals in hibernation would strain life-support systems about as much as one individual awake and active. If every individual spent one year as crew and nine in hibernation during the journey to Alpha Centauri, that 150-year journey suddenly becomes a fifteen-year journey, which is far more doable within a single crew's lifespan.

Nutrition, Waste, and Green Living

Colony ships and colonies will need to provide renewable foodstuffs to meet nutritional needs without the requirement to carry masses of stored food. Aquaponics have long been theorized to be important in interstellar travel and early planetary colonization, and with good reason: They handle both food production and waste conversion. Modern aquaponic systems involve bacteria, fish, and plants in a mutualistic system where fish provide nutrition to plants in the form of urea, bacteria break down solid waste into inorganic sediment, and plant roots feed the fish, with the leaves, fruits and a certain percentage of fish available for harvest.

Anthroponics takes this one step further, processing human urine into urea as well. Human manure is suitable as a soil fertilizer with one major drawback: It is up to fifty percent bacteria by dry mass. Sterilization processes exist, but as anyone who's read about *E. coli* outbreaks from lettuce or organic produce can attest, they are not perfect.

Meeting on-board nutritional needs of colony-ship residents will be a key logistical challenge. Even now nutritional guidelines are constantly being modified and the relative importance of various micronutrients continues to be discussed. The role of carbohydrates and fats in health and disease (what's good, what's bad, and how much) continues to be hotly debated. In the recent past, vitamin D was thought a low-yield as a supplementation target and omega-3s continue to prove ever more crucial to an increasing number of physiologic processes. Additionally, we continue to learn more about trace metals as both poisons and piconutrients.

Precious Metals

One rarely explored issue, the depletion of pico- and micronutrients on long interstellar voyages, is a particular area of concern. These include such metals as copper and zinc that are more often thought of as ingredients in brass than important cofactors for the functioning of enzymes. Though we require mere micrograms of these nutrients per day, deficiencies can lead to fatal disease. A well-developed aquaponics system could address this by storage of dry minerals, chelation with amino acids to improve solubility, and dilution into the aquaponics system. Many seaweeds, including commonly eaten varieties such as porphyra, chlorella, and spirulina, concentrate important minerals including iodine, zinc, gold, and copper efficiently, and would be easily accommodated by a sufficiently sized aquaponics system.

Lead Poisoning and Acid Air

A final concern regarding waste and nutrition is heavy metal poisoning. Ships will be composed of many metals with toxic potential, and plastics, coolants, fuels, and other synthetic fluids will inevitably off-gas volatile organic compounds. The latter was important enough to NASA in their quest for long-term journeys that they investigated plants' abilities to absorb and neutralize these organic compounds, the results now seen in any garden store with multiple common and hardy houseplants proudly marketed as air-purifying.

Heavy-metal toxicity is a more difficult problem, but one that has a technical solution. Because heavy metals will likely be constantly introduced into the biome, a constantly filtering system will be important. Complex mammals and most plants have systems to deal with this known as sequestration. However, the modern technological world full of metal and alloys contains concentrations well beyond the body's capacity to cope; simply walling off toxic metals within the body can wreak havoc as the metals continue to accumulate. The effects of lead toxicity on the brain are well known from the poisoning of children by lead-based paints prior to the paint ban a few decades ago; likewise, aluminum toxicity plays a known role in dialysis dementia and has been implicated as a risk factor in several cancers and neurodegenerative diseases.

The tendency of shipboard-grown plants to accumulate metals will likely exacerbate the problem. Genetic engineering, nanotechnology, and simple physical chemistry offer potential for solutions. Multiple bacteria, simple multicellular organisms, and plant species have an alternate strategy for dealing with heavy metals: protein channels known as efflux pumps, which simply dump the toxic substances back out of the cell. These genes may prove good targets for integration into humans and would likely to continue to benefit us once we colonize. Another strategy involves engineering sacrificial plants that preferentially sequester toxic heavy metals from the aquaponic system, which can then be disposed of. Nanotechnology designed to scavenge toxic metals from our bodies, rendering them non-bioavailable, is an alternate strategy. Organic compounds known as chelators bind to metals rendering them inert and destined for elimination in urine or feces. Such compounds already exist and are effective in cases like lead poisoning. Materials science innovations may also play a role with surfaces that preferentially bind toxic metals suited for submersion in aquaponics systems.

Vitamins

The hominid lineage stands in an odd spot, born of committed frugivore primate, and now functionally obligate omnivores, and as a result we have an odd array of nutritional dependence compared to other animals. Many substances we call vitamins or micronutrients are things other animals produce independently. Vitamin C, for instance, is made endogenously by most mammals. Its production enzyme was lost somewhere in early primates without consequence because it was unnecessary to animals that got so much of it from their diet. Vitamin K, on the other hand, important for blood clotting, skin, and neurologic function, is something most mammals with carnivorous diets make endogenously, given their low leafy green intake.

The genes and proteins that make many, if not most, of our vitamins are found in other animals, often mammals, or in the bacteria that line their guts. Any of these would provide profitable avenues of manipulation of genetics, ones that would continue to pay dividends as we colonize new worlds.

Fitness

Maintaining fitness on board is incontrovertibly important and has been a concern of NASA and other space agencies since the 1960s, when they first noted that rapid deconditioning of astronauts in microgravity can occur in a matter of days. Since then, a variety of exercise equipment and conditioning regimens have been developed and continue to be refined. While some health concerns associated with zero gravity would be obviated under the gravity associated with thrust, others would remain, particularly the need to maintain cardiovascular and muscular health on board an undoubtedly crowded ship.

Prior to industrialization, physical fitness was closely linked to participation in tasks and sports, meaning a certain level of skill was required for maintaining fitness. Whether throwing a ball, executing a well-timed jump, or the complex coordination of movements of wrestling, dance, or gymnastics, the integration of the body, brain, and skill cannot be overlooked. For thousands of years, sports have been instrumental in fitness and should likely continue on our colony ships, which would necessitate one or more areas large and empty enough for group participation. This might initially seem an extravagant luxury, but its justification is clear.

Infectious Disease

Hygiene and sterility can only go so far in preventing the spread of disease on board a ship. This is because many pathogens— especially bacteria—are considered *normal flora*, meaning they're regular hitchhikers on our body. Some we simply couldn't live without (like probiotics). Others we simply can't be rid of. The potential for pathogenesis, then, is one we will always carry, and, with a small population using a closed system for waste and nutrition, infection is more an inevitability than a risk.

Aggressive infection control will thus be paramount and will likely require quarantine spaces with self-contained life-support systems. Normal and pathogenic flora quickly adapt to traditional antibiotics, the former often "helpfully" transferring those genes to new pathogens. Molecular techniques already play a role with rapid drug development tailored to antigens (identifying structures on pathogens' surfaces) and vat-grown monoclonal antibodies to specific microbes and the toxins they produce. Immune-boosting drugs are likewise being developed. These development cycles are currently on the order of months to years, but the ability to do so "on the go" and in close to real time as new pathogens present themselves may be the difference between life or death for the crew of our starships.

Novel technology may shine in this realm. While there are certainly places for nonpathologic bacteria to live in our bodies, there are certain areas where no bacteria should be, such as blood, brain, lungs, and kidneys. Artificial cells or nanomachines with a zero-tolerance policy for bacteria that restrict themselves to these tissues could limit the pathogenicity of any bacterium.

Reproduction and Child Rearing

Shipboard reproduction is likely to be a functional or obligate necessity given the lengths of our journey to the stars. Beyond simple pragmatics there are specific concerns of any child born on board a ship of limited size and population. Next, how are the children selected? Will crew be free to procreate on their own (see Kacey Ezell and Philip Wohlrab's story, "Stella Infantes," and Dan Hoyt's story, "Exodus"), or will embryos be preselected through exhaustive genetic screening? Will they belong biologically to one or two crew members or derive from preserved Earth sperm and ova in order to preserve genetic diversity? As discussed, many negative traits can be out-screened but positive ones are simply too complicated to select for with any real predictability, never mind the social implications.

Even before involving the consequences of genetic engineering, we must contend with the fact that few desired traits "breed true." Heritability for intelligence is pegged at about 0.7; thirty percent of the variation lies outside of genetics. On average, the children would likely have a greater range than the parents. In other words, the least intelligent child will be substantially less intelligent than the least intelligent in the parents' generation—an example of "regression to the mean." On the other hand, you can have too much of a good thing. There is now genetic evidence to support the assertion that, genetically speaking, the old adage about a thin line between genius and madness holds true, at least regarding such illnesses as ADHD, bipolar disorder, and autism spectrum disorder. There appears to be an optimum number of "good" alleles with respect to intelligence, with too many being just as problematic as too few. Similar properties apply to the immune system and physical robustness.

In a crew selected for exceptionalism, each subsequent generation will trend back toward human baseline. Those who become colonists, potentially several generations downstream, may be far less capable than their predecessors, even with careful manipulation of the factors within our control.

Regarding reproduction, artificial wombs would be an incredible advancement, but they ignore the dyadic nature of the biopsychology of mother and infant. This involves complex hormonal interplay between the two which affects the infant's development and prepares the mother to become the primary attachment figure. On the day of birth, a newborn can identify its mother's voice as distinct and calms more quickly to her touch than others. The mother produces milk tailor-made to her baby's immune system and neurodevelopmental needs. The role of pregnancy in the attachment process is such that even when adopted at birth to good homes, adoptive children's outcomes lag. Attachment figures are important. For infants and children, the ability to identify a primary caregiver, someone whose primary responsibility is their safety and emotional well-being, relates to a range of positive outcomes, from emotional and physical resilience to intelligence. Absence of a primary caregiver likewise has striking effects.

Children need enriching environments, full of things to explore and learn about and to facilitate self-exploration. This requires space and supervision. Children also need varied environmental contaminant exposure, which could be challenging or otherwise problematic on a spacecraft. The "hygiene hypothesis" states that healthy immune functioning and low allergic tendencies are linked to dirt exposure, different foods, and a range of pathogens, without which children are at greater risk of allergic response to alien environments or immunoincompetence to mutated or xenopathogens.

Psychology

The basic personality structure of an ideal colonist is clear. In daily life they must be measured, agreeable, conscientious, and tolerant of boredom. Yet on longer scales, they must stomach risk, think adaptively and problem solve in the moment, and be self-reliant. However, no matter how carefully our early explorers are screened, generational timescales mean further thought is required.

Mental illness will occur aboard the colony ships. Much will have nothing to do with the setting and everything to do with mental illness being rather common, with depression, anxiety, and substance abuse topping the list. Some may develop PTSD from disasters and near-misses. But some conditions may be unique to shipboard life. From studies of remote posts, submarines, and long-term space missions, as well as animals in captivity, we know something of what we are likely to see during interstellar journey. Obsessive compulsive disorder will likely be over-represented. A life bound by and dependent upon routines will inevitably lead to the performance of routines without clear purpose. The fact that so many shipboard routines are matters of life and death will lead to the belief that these nonsense routines are similarly imbued. Luckily, treatment is straightforward, if not always easy.

The highly controlled environment may also lead to sensory processing disorder, in which the normal unconscious filtering of stimuli becomes impaired. Light sources that don't match the characteristics of shipboard lights will be jarring and even head-ache inducing. Sounds not normally heard, hums of a different pitch or intensity, will sound like nails across a chalkboard. This can be managed by carefully varying our background noises and lights in a way that keeps these filtering pathways active.

Humans can set aside personality differences to get extraordinary jobs accomplished. Natural disasters, heroic projects (e.g. space missions) and harsh environmental colonies (e.g. Antarctic bases) unify people who normally wouldn't give each other the time of day. But there are people who do not or will not continue to put the mission first. They might evade psychological screening, which remains an imperfect "science." Problems might first occur in children of the original explorers, beyond the careful control of pre-mission planning. Like a speck of grit against a bearing, small differences can magnify, insults and friction growing larger. Humans, for all our vaunted rationality, can reach points at which emotions and relationships cannot be saved by logic, facts, or by the dire consequences of irreconcilable differences or antisocial behaviors. It might not even be intentional, just an inability to function at the level demanded of the closed, high-stakes system of exploration and colonization.

There will be misfits. There will be malcontents. And we will not always be able to find a place for them. They cannot leave for different environs and different company; they cannot be failed out of basic training or graduate school; they cannot be incarcerated or otherwise sequestered. The handwavium of deep freeze may yet come to fruition, obviating this concern for the time. But our ships and colonies will have to deal with this. There is no banishment in the harshness of vacuum or the alien wilds of a distant planet. There will be people for whom there is no rehabilitation; worse, some may not be truly criminal, but simply incapable of meeting the standard of functioning as early explorers.

COLONIZATION

Colonists' transition from shipboard to alien planets will mark one of the biggest moments in human history, but also will be one of our most vulnerable. The ship, while associated with its own challenges, is a highly controlled environment. A new planet hosts novel challenges: new environmental variables, from gravity to climate to light spectra and diurnal periods, all in addition to the known risks of any major infrastructural undertaking or frontier-breaking. No matter how much automation, or how sophisticated defensive measures become, there will always be an element of human risk if humans are present.

Chronobiology

We take days, nights and the passage of seasons for granted. Far from being incidental, our biology is indelibly linked to the passage of time, with cycles upon cycles dependent on environmental factors.

Though easily taken for granted, adaptations to twenty-four-hour days are intimately connected to immune and endocrine functioning. Some of us are night owls, others early risers, but all of us function best on a twenty-four-hour sleep-wake cycle, polyphasic experimenters notwithstanding. Studies have proven that even in total isolation from external influence (such as clocks or sunlight) our bodies approximate a twenty-five-hour rhythm of sleeping and waking, with the extra hour thought to provide a physiologic buffer. Most are familiar with melatonin, a sleep-regulating hormone. Many are also aware that blue LEDs especially, but all lights generally, can disrupt restful sleep, partly owing to interference with hormonal determinants of sleeping and waking.

The length of a single sleep "cycle" from shallow to deep to REM is approximately ninety minutes. And as it turns out, each of these cycles is important for growth and recovery. Deep sleep is where most of the "rebuilding" comes in. This is when our stress hormones are at their lowest, and our immune, bone, and muscle cells undergo the most recovery as growth hormone gets released in pulses. During the mid-range of sleep (including REM sleep) neurologic structures consolidate the day's changes and refresh themselves, which is vital for optimal cognitive and emotional functioning.

Ideally, we get five full cycles of sleep a night, some more, some fewer. What would happen on a planet with a shorter average day where we operated on fewer consecutive cycles of recovery before the stresses of the next day? How about a planet with a longer average day? This question remains unanswered, but effects on immune function, stress response, longevity, and cognitive functioning could be expected.

While day-night cycles would occur on any planet, seasons as we know them may differ considerably with respect to variation in climate and starshine, from minimal on planets with low eccentricity and minimal deviation of rotational axis from the ecliptic, to far more drastic on other planets. Latitude of the colony would also play a role.

Though incompletely understood, there are known impacts on seasonality and health. Certain pathogens like the influenza virus have a clearly seasonal pattern. The risk of certain disorders, including such disparate illnesses as schizophrenia and autoimmune diseases, are associated with being born in wintertime, even in our climate-controlled and artificially illuminated world.

Environmental Considerations

From air composition to the makeup of soil beneath our feet, alien worlds present the possibility of environmental toxicity. As previously discussed, these include gaseous compounds, such as carbon monoxide or methane. Even oxygen can be toxic if it is too highly concentrated. Beyond hermetic habitats and environmental suits, genetic engineering may prove useful here too, as many other animal species are equipped to tolerate broader ranges of atmospheric conditions.

Here again, trace metals may be problematic, either through excess or absence depending on biological significance and concentration. (Selenium, for example is essential yet toxic at high amounts.)

Xenobiology

Complex life—or even simple life—has yet to be conclusively identified out amongst the stars, but it remains plausible, regardless of likelihood, that any world worth colonizing might have life of its own with which to contend.

Some xenobiological precautions are obvious. For example, avoid being eaten by local fauna or getting injected with venom via stinger, tooth, or claw. Then there are seemingly innocuous things like touch, which anyone who has endured a jellyfish sting or the itch of poison ivy can attest. Similar experiences on an alien planet could range from uncomfortable to lethal. And some plants are best left uneaten.

We will likely need a multipronged approach to manage exposures to biologically derived poisons, venoms, and toxic compounds. Appropriate quarantine procedures would be needed in the early stages of colonization, but as colonies mature and integrate with their new environment, the blending and eventual loss of boundaries is inevitable, and perhaps hoped for.

Preventing overwhelming, potentially fatal allergic/immune responses to nontoxic substances would be important since allergic response is often based on inadequate exposure to specific potential allergens early in life (see Cathe Smith's story "The Smallest of Things"). Inhibiting cells implicated in allergic response is a nonviable solution, as those same cells are crucial players in wound repair and defense against multicellular parasites. We can teach our immune systems to ignore benign allergens through immunotherapy, although this process currently takes months to years. A more rapid system involving direct epigenetic modification of the cells that eventually give rise to allergic response might be possible in lab settings via artificial cells specifically designed to upregulate this response, or through direct gene editing of specific immune genes on a case-by-case basis.

Poisons and venoms, unlike allergens, cause direct physical harm through numerous processes which can include inflammation cascades or clotting pathways (both of which are seen in spider and snake bites). Many plant-derived poisons work by binding proteins responsible for biochemical functioning, rendering them inert. For toxins that work via cascade induction (e.g. rattlesnake bites), antibodies specifically targeting the individual compounds have been successful. Antidotes to enzymatic poisons, on the other hand, work by degrading the poison directly, or literally pushing it off the protein to which it is bound. These strategies are too toxin-specific for generalized application. Neither can be made prior to the existence of a known toxin.

If the panspermia hypothesis is correct—if cellular life on other planets is similar enough to our own, or if they are simply rugged and versatile enough—potential for xenoinfection likewise exists. Viruses rely heavily on a cell's own "machinery" to continue their infectious cycles. The likelihood of a xenovirus, even with reliance on DNA, RNA, and similar amino acids, to effectively hijack a cell that uses different genetic code is biologically improbable. Bacteria, conversely, simply digest and reproduce, regardless of environment (be it a hydrothermal vent or within brain tissue). The damage bacterial infection causes is secondary to this. Thus, xenobacteria, or whatever single-celled alien analogs may exist, if tenacious enough, pose real risks.

What these situations share is that any technical solution must inevitably be reactive, rather than proactive, by nature. Biological mechanisms to prevent susceptibility require either hardcoded genes, or the adaptive immunity of exposure and response (the mechanism behind vaccines). So how do we react to unknown biological risks without major consequence at every turn? That first small step could be someone else's last, as would every individual's after for untold thousands of steps (see Sarah Hoyt's story "Burn the Boats"). Science offers a tantalizing, if morbid, solution in the form of chimeras. Like the mythological creature of legend composed of parts of disparate animals, the scientific, modern chimera is a cell or other organism containing DNA of two species. We use simple chimeras in several important medical therapies. Bacteria modified to contain the human insulin gene allow mass-production of medical insulin, keeping diabetics alive. Rituximab, an artificial antibody produced by mice transfected with human genes, treats certain cancers and autoimmune diseases. And, for the last thirty years, we have used human/mouse, human/pig, and human/monkey chimeras in biomedical research.

We could test our environmental exposure risks in a more sophisticated version of the canary in the coal mine, although mice or another mammal would be a better substrate for chimerism. Ideally, such chimeras would have human immune systems, detoxification systems (livers and kidneys), and digestive intricacies and limitations in a smaller, low-maintenance, and ethically only marginally problematic package. These chimeras could be exposed to pathogens one by one, with each exposure informing a new genetic or protein modification to develop.

The task is daunting, but the seeds of technical solutions lie within our grasp.

Colony Growth and Mental Health

The children of *Homo stellaris* will be protected as much as we are able. We will do our best to give them safety and belonging along with freedom to explore themselves and the world. But that selfsame world is more dire than the one their parents left, and in order to meet the challenges of their new worlds, become inured to them, and build their place, our children will inevitably have to face the threats.

The stories we once told children echoed the world around us and were intended to help prepare them for the things they saw glimpses of with little understanding: from sins to struggles to tragic deaths. Nothing can change the nature of pain or of loss. The original children's fables and fairy tales have been remarkably sanitized in their current pop culture incarnations. Historically, these stories were full of violence and greed and death. Good guys won because they were stronger or smarter. And lost when they were arrogant, were betrayed, or were simply unlucky. These things will always hurt. Yet the attitude with which we face them can make all the difference.

Global mental health research has yielded some interesting

findings: The rates of depression and anxiety are rising in the West and are considerably lower in Asia and Africa. Furthermore, as Asia and Africa modernize, *their* rates in turn grow. The rate of psychopathology in many ways has an inverse link to the safety and standard of living of the culture in which an individual resides. While at first counterintuitive, the argument is threefold:

1. The modern world carries with it an expectation of safety and comfort, and the failure of this unattainable ideal causes pain.

2. Much of our vocational time is spent avoiding negative consequences instead of attaining rewards.

3. The typical work day in the modern world leaves behind little sense of accomplishment.

These are all things that can and will be different for our descendants on distant planets. Their lives will in many ways be harder than ours. But we can make sure that they take pride in their work and in each other. They can grieve their losses not out of a sense of unfairness, but in recognition that life is all the more precious for it. They can marvel under old stars in new constellations. We can direct them toward a new culture, not so different from the old one, in which risk and pain are the accepted costs of creating a new future.

NEW WAVES OF IMMIGRATION

Colonies are likely to be founded by a few hundred individuals. Certain traits might be more common in these groups. Individuals with rare or novel mutations are likely to be over- or under-represented. And those rare mutations are more likely to take over and dominate the genetic makeup of the population during the early period of rapid expansion. This could lead to a population of colonists that is more homogenous or overrepresented in certain aspects, which may or may not be of any relevance. One could imagine a colony in which red hair predominated, or curly hair, so that they viewed those without such otherwise relatively unimportant attributes with suspicion.

Genetic Bottlenecks

Genetic drift occurs when a population is isolated from the rest of the species—most often by geography. The isolated population will begin to change independent of the other population. Some of these shifts will be adaptive in nature, in response to the different conditions they face. This will no doubt be augmented by our colonists' vital need to tinker with their own biological functioning and may lead to a situation in which they contain alleles, or, in fact, whole genes that the Earth population does not have.

Our colonists will have spent generations out of contact with the rest of humanity. For context, our first colony ships are not likely to leave within the next century, and any journey they take will last at least another century, if not many more. In the last century, change has been incalculable on many fronts. In essence, the colonists' separation may rival the length of time between now and when Shakespeare's first play was performed on the banks of the Thames, or even the fall of Rome.

Imagine the first colony ship to land on a planet circling the Alpha Centauri system, setting up base camp and signaling home, after a few centuries, that they will be ready for more colonists. The next colony ship arrives the better part of two centuries later: The initial colonists and those that follow will have been separated by time, by science, and by biology. The initial colonists will likely have undertaken numerous genomic editing projects, which may not be the same solutions the new colonists prefer. Those new colonists would have experienced scientific innovations and breakthroughs on Earth and may have unrecognizably different genetic baselines or otherwise have faced the same problems with different solutions (see William Ledbetter's story "Bridging").

Infection

And then we have the fact that pathogens will have continued to evolve on Earth, and humans will continue to respond to them, often without technological or medical notice (after all, who goes to their physician for a common cold?) while alien pathogens will have done the same amongst the colonists. Two different cold wars, two different responses, and ultimately two very different looking pathogens that were indistinguishable centuries prior.

As it stands in the current day, individuals from the same general geographic area at the same time with certain disease compromising the immune system are told never to be in the same room as each other (cystic fibrosis). Their microenvironments result in pathogen mutations that mean the same bacterial population that lives peacefully in one individual's lungs can quickly become an insurmountable infection in another individual's. A hospital three hundred miles away from another will note that the same diseases are susceptible to different antibiotics. These differences will be even greater across stars and centuries. And a prospective mechanism of avoidance is not clear. It is not unthinkable to imagine first contact between two groups of colonists involving the timid introduction of comingling rodents from each habitat as they pare away, death by death, at the potentially deadly infections they harbor.

Phenotypic Change

Phenotype, or physical characteristic, is a combination of genetics, environment, our choice in how we interact with environment, and even our parents' interactions with the environment. Human genotype has changed in few ways in ten thousand years, while phenotype has changed and been more variable far more drastically. From average height, to the shape of the nose, to our ability to digest or tolerate various compounds, these changes can occur with astonishing rapidity.

What changes might colonists see over just a few generations? Changes in skeletal mass or structure go almost without saying. Average human height and skeletal robustness respond quickly to food source, modes of living, and other factors even within relatively stable lineages. Pre-agricultural *Homo sapiens* was remarkably tall and muscular on average, standing five foot ten inches and approximately two hundred pounds for males, which by the early to late agricultural periods had decreased to five foot six inches and approximately one hundred forty pounds, then began increasing sharply in height but not skeletal robustness during the industrial period. Cranial volumes (and thus brain size) were larger for primitive hominids than early industrial hominids. Interestingly, there is little evidence to support that any of these major changes had much to do with genetic mutation.

Reproduction

True genetic incompatibility is thankfully rare amongst humans even between areas with exceedingly low gene flow. Even populations that have been isolated from others for thousands of years can generally intermingle with little consequence. However, when we begin tinkering with our own DNA the prospect of incompatibility becomes far greater. A peculiar aspect of genetics is the concept of pleiotropy: that one gene, through the different ways it can be spliced and read, can produce multiple proteins, each of which can have multiple activities on multiple cell tissues. This multiplicative effect means that if we take one population that has edited one gene, and one population that has edited another, the effects might be somewhere between difficult and impossible to predict. The results might be simple infertility, or they might be deleterious consequences to the progeny.

INTERSTELLAR REINTEGRATION

The goal of colonizing the stars, one would hope, is that our daughter colonies, someday far in the future, will reestablish communication, if not physical interchange of material and individuals. Each colony ship will have left the planet at a different time in history, with a small group of people for whom certain traits previously deemed unimportant will be more or less represented, and who have chosen a different set of biological and technological solutions to interstellar travel.

These ships will make landfall on distant planets, with different challenges, and in turn choose different responses and solutions. Some consciously and with forethought, some instinctively, and some dictated by passive biological response. For all our differences in language and physical characteristics, *Homo sapiens* has remained remarkably homogenous, with even tribes and ethnicities separated for thousands of years from others able to reintegrate or at least communicate rather easily. That said, we are also remarkably different, in language, in culture, and especially in biology, with a thousand solutions to a thousand different environmental challenges upon our planet. Flung across a hundred stars, on a thousand planets as different from each other as they are from Earth, with fantastic yet plausible technologies, our children will change in ways that cannot be fully anticipated, some forced by their environments, some through random chance, and some through deliberate remolding of their own clay (see Todd McCaffrey's story "At the Bottom of the White" and Brent Roeder's story "Pageants of Humanity").

Reuniting will no doubt be challenging. They may find their microbiology or physiology initially incompatible. They may look at their sister colonists' differently shaped and colored bodies and not recognize their own kinship. They may find the children born of parents from two different worlds unsuitable for either. But we look to the future with hope, that those who solved so many problems will solve these too.

A FUTURE, IF YOU CAN KEEP IT

The psychosocial challenges in our reach for the stars will be manifold. The unrelenting *sameness* of colony-ship travel will place immense demands on individuals' resilience and optimism, while straining the ability of any crew to maintain cohesion over timescales measured in decades. Familiarity, after all, does breed contempt. Perhaps even more so when high-achieving individuals from different pathways and different perspectives are forced to work together. There will be unavoidable friction between individuals, which if grown to the level of groups or factions, could lead to total failure of the ship and thus the mission.

The children born between the stars may not be able to adapt to new planets. And those born on new planets will find their early years marked by more demands, more stress, and sadly more death, yet also more opportunity and even meaning than those left behind. Imbuing them with the resilience and skills to handle this brave new world without depriving them of the innocence of youth will be a careful balancing act.

Humans are not inherently communal—a shared goal is not enough to keep us acting in concert for very long. Yet humans maintain the most complex and persistent cooperative networks of any animal. We are at our best as individuals, working together, bonded by our journey, by the work and the experiences we share, and by the fact that none of us can reach the stars without the others. Reciprocal altruism has always been the way living creatures work best together, and that will be no different in the future.

Almost two million years ago, a small tribe of *Homo ergaster* left Africa for unknown environs, and several small groups followed, in different directions, likely never seeing each other again. In doing so they opened up not only the rest of the planet to their descendants but created the fertile substrate for change in which *Homo sapiens* evolved. As the first colonizing crews leave the heliopause, perhaps they will create the spaces in which *Homo stellaris* will evolve.

Exodus

Daniel M. Hoyt

Daniel M. Hoyt is a systems architect for rocket trajectory software, a professional SF/F author, and an expert on royalty calculations for indie presses. Since his first sale to Analog, Dan has appeared in other magazines and anthologies, notably Transhuman, all three of the Baen Suburban Fantasies edited by Esther Friesner, and Mike Brotherton's groundbreaking *Diamonds in the Sky*. Dan has also edited two anthologies, *Fate Fantastic* and *Better Off Undead*. He is working on sequels to his debut space opera novel, *Ninth Euclid's Prince*, and a collaboration with his wife, Sarah A. Hoyt. Someday, he may even update www.danielmhoyt.com.

"When you said you wanted to go on your junior year abroad, we thought you meant France, not the Moon!" Virginia Grant's mother, Helen, cried. Her normally youthful, sun-kissed complexion darkened, revealing her true age in a flash. Her husband, Aaron, nodded and puckered his lips as if he'd just bitten into a bitter lemon, but said nothing.

Ginny sighed. "I said I was going on a *lunar* year abroad."

"But *why*?" Helen wiped away her tears dramatically to punctuate her distress. "Why the *Moon*?"

"You always hear what you want, not what I say, Mom. I've *always* wanted to go to space. I made no secret of this, ever. But that doesn't fit with your perfect-daughter strawwoman." Ginny hauled her heavy backpack off the floor at her feet, grunting. "I knew this would be a waste of time." She plopped the backpack into her lap, waiting for the next phase of the argument: Dad logic.

"Now listen here, young lady," Aaron chimed in, as if on cue. "We didn't pay all that money for you to get three degrees just to flush them down a toilet. We still don't understand why you didn't want a useful degree, like Sociology or Art History or even Political Science, but we supported your decision to go into STEM, didn't we? Even when you added new more useless majors?"

"Yes, Dad."

"What would you even *do* with biology in space?"

"You're kidding, right? You've seen all the documentaries on *The Exodus*, haven't you? The colony starship they want to launch in two years?"

Helen sniffed. "Of course, dear. Is that what this is all about? You want to get away from us so badly, you're willing to go to, to, wherever it was, to... *out there*?"

"Proxima Centauri b. That's where *The Exodus* is going. And, no, Mom, I'm not trying to go with them. There's still a lot of science fiction that needs to become actual science before the launch can happen. It will probably get delayed a few more times while the technology catches up to the vision. The passenger list has been filled for the better part of a decade, anyway, so I wouldn't be able to go, even I wanted to." Ginny struggled to her feet. "But I want to *help*. I want to be there for the launch, knowing I'd made a difference, done *something* that helped them get there. I've been working toward this since they announced the project when I was thirteen." She spun on her heel and stomped to the door.

"Now listen here, young lady," her father called, but she didn't hear anything more after slamming the door behind her.

Ginny chucked her backpack into her Tesla Model H Anniversary Edition and settled into the passenger compartment, scrabbling for a tissue. Salty streaks staining her face, she swiped at them and blew her nose into the now damp tissue. A mist of antibacterial orange burst from the roof-mounted vents, settling over her.

"Would you like to go home, Ms. Grant?" The Tesla auto-navigator waited patiently for an answer.

"Home, James."

Her Tesla ascended into the sky traffic smoothly, as always, while Ginny fumed.

What did she have to do to get it across to her parents that she could never be happy staying on Earth, not after *The Exodus* had been announced? It was *space*, wasn't it? It had been more than a century since the Golden Age of Science Fiction introduced humans to the possibility of space travel in an earnest, concrete way, and humanity embraced the concept with open arms. *Ginny* had embraced the concept, too, with more than just her arms. Her mind, her soul, her very *being* was imbued with the quest. Her parents just didn't see it through their self-imposed blinders.

That first announcement was *it* for her, the missing piece of the puzzle that was Ginny. She knew what she wanted to do with her life from that moment on. The Jacobsen–White variation of the Alcubierre warp drive appeared to *work*, at least on a model scale, stoking a fever for anything interstellar and laying the foundation for a legitimate star-colony program, a fever that burned hot even after the results of the JAWW drive testing was revealed to be inconclusive. The star-colony program didn't fall apart, as the Fix Earth First organization predicted, it just retooled for the technology that *could* be used for an interstellar voyage—a relatively slow stroll to Alpha Centauri at one percent the speed of light, for a voyage of about 425 years. Two years later, *The Exodus* was announced, with those limits in mind— along with a widespread hope that those limits could be changed by the time of the actual launch.

By then, Ginny was all-in for space, deftly shifting the focus of her classes science-ward, abandoning the arts curriculum her parents had been pushing on her since birth. Fervent Feffers—as the press dubbed members of the Fix Earth First Society—for longer than she'd been alive, the Grants had insisted that Ginny learn their worldview first, before any others, and understand the dangers humanity posed to the Earth. It wasn't that Ginny didn't agree with any of the Feffer propaganda—some of it was based in actual science, but some was no better than pseudoscience extrapolated from what amounted to shock-inducing projections from drug-induced visions by nonscientists. Ginny just couldn't get on board with the Feffers' stated goal, Fix Earth First. The Feffers were committed to opposing anything relating to space, because the Earth wasn't yet *perfect*, by whatever measure the

Feffers were using, and humanity had no *right* to go to space while the Earth was still in need of Feffer assistance.

During Ginny's formative years, along with the Feffer agenda, she'd managed to study history. Apparently, the Feffers had reached different conclusions than she, since she saw the imperial expansion of civilizations like Rome and China as the *cures* for their sicknesses, rather than the *causes*. Expansion brought new knowledge, new ways of thought, new technologies, new techniques for coping with the problems they experienced at home. If humanity wasn't getting along well with Earth, then it was time to expand beyond Earth's boundaries, think outside the dome, and maybe figure out how to fix that broken relationship.

For Ginny, the only answer for the questions raised by Feffers was the stars.

"Arriving home, Miss Grant."

Ginny sighed. At least the journey home from her parents didn't take 425 years.

"Ginny, you okay?" Robert asked over the v-comm while munching on an unnaturally bright orange bioengineered carrot. Worry lines creased his upraised brow, his soft chocolate brown eyes opened wide with alarm. He'd known Ginny since she was an eight-year-old know-it-all, and nothing could ever ruin their fierce friendship.

Ginny brushed aside long wisps of strawberry blonde bangs and forced a smile. "I'm okay, Rob, really. I just had a little argument with Mom and Dad."

Robert grinned. "So, Tuesday."

"Pretty much, just ten times worse. Mom broke down in crocodile tears, and I left. I just got home."

"What happened? Did you tell them you're adding an aeronautical engineering minor?"

Ginny smirked around her sniffles. "Not yet. I told them I applied for a fellowship on the Moon."

Robert whistled. "That must have been fun. No wonder they lost their minds. I take it the Feffers would prefer you stay closer to home? Canada? England? Australia?"

"France, to be exact." She sighed. "At least they figured I'd go somewhere English isn't spoken. At least not English as we know it."

Robert shook his head. "Yeah, Moonish is trippy. It's like a mashup of tech acronyms, Latin and some Greek, all glued together with English syntactical construction. *Gyrizo lavi PVC valvae*."

"*Anglicus triumphos, filha da puta!*"

"Wait, was that *Portuguese*? No fair, you brat!" Robert blinked as the words sank in. "*What* did you say about my mother?"

Ginny giggled. "I was thinking of my mother, actually."

"Well, that's all right then." He waggled a finger at her mock-menacingly. "But don't say anything like that again, young lady, or I'll take you over my knee and spank you."

"You wish." Ginny immediately wished she'd kept her mouth shut; the conversation stopped abruptly after the mildly salacious exchange. Ginny knew Robert *liked* her, but he didn't share her passion for space, so a romantic relationship was out of the question. She mentally kicked herself for encouraging his inappropriate thoughts.

Robert cleared his throat. "Seriously, Gin, you're actually going?"

"I hope so. I'll find out next week."

"You know I'll miss you."

"Yeah, of course I do. I'll miss you, too. You're my bestest friend; always will be, even if I'm on the Moon without you."

"I'll only be a v-call away, Rob. You know that."

"I know. Still—"

"And I might not even get in," Ginny added quickly, before he could dig another romantic hole for himself.

"You'll get in, Red, you know you will." Robert ticked off her majors on his fingers. "Biology, chemistry, physics, and now aero engineering." He paused. "I feel like such a slacker with just mechanical and electrical engineering. Why *wouldn't* they want you? They still need to figure out cryonics for a voyage that long, and your skillset is *perfect* for that. If *anyone* can figure it out, it's you."

"I'm still just a student, Rob. There's probably some Nobel laureate out there that will figure it out before I ever get a chance even to work on it."

"You'll get it, I promise. I believe in you, even if you don't."

Ginny blushed. "Thanks, Rob. You always make me feel better after an argument with my parents."

"That's what . . . *friends* are for, Gin. You can always count on me."

The slight pause and emphasis on the word "friends" wasn't lost on Ginny, but she stuck to her no-romance rule. "I know, Rob. Gotta go now." She broke the v-comm before Rob could see her tear up.

An excruciatingly five-day wait later, Ginny had an answer on the fellowship. When her v-mail announced the v-message, she stopped julienning some bio-carrots so abruptly she nearly slashed her hand. Shaking, she swallowed back bile, dropped everything onto the cutting board and dug her fingernails into her palms.

"Read vee mail from Lunar Fellows."

"Congratulations, Virginia Grant. You have been accepted for the Preparation Team on Lunar Base C, pending skills and language assessments and completion of physical preparation courses. If you wish to accept, please respond to the Prep Team within the next three days. Again, congratulations, *neosyllektos*."

Ginny took a deep breath and tried hard to calm her nerves. "Call Mom."

The v-comm zizzed to life, and a hologram of her mother's upper body and head appeared over the island stove.

"Hello, dear," Helen sniffed. "Are you done ignoring us now? We've left several messages."

"I got in, Mom. I'm going to the Moon."

Helen's mouth formed an "O" and her face crumpled, as if she'd just been stabbed in the back. "Right now?"

"No, Mom, not right now. Next summer. I have to train all school year so I can pass the physical exams. I could do the Moon-ish exam tomorrow, if I needed to; I've been studying it for five years already. And I know I'll pass the skills assessment." Ginny smiled. "Especially with my new classes in aerospace engineering."

"Your what?" Helen looked thunderstruck. "Aaron!" she called off-screen. "AARON!" She looked back at her daughter murderously. "You *added* more useless STEM classes? Without even telling us? AARON!"

Her father's holo-torso muscled into view, partially pushing his wife's view out of the v-field. "What's this I hear about more STEM, young lady? Why? When does it stop?"

"I needed to add them, Dad. I'll need them on the Moon."

"Good lord, *that* nonsense again?"

"Yes, Dad. The Moon nonsense. It's real; I've been accepted."

Aaron Grant opened his mouth, but no sound came out. He closed it again and opened it once more, with the same result. He closed his mouth and pivoted to his wife, muttering just one word. "Helen?"

"And she's taking more STEM classes!"

"I don't care about that. She's going *away*! *Forever*! And you're worried about some *classes*?"

"I'm not going away forever, Dad. It's just a year. Lunar year abroad, remember?"

Her mother's lips compressed into a straight line. "It might as well be forever."

Ginny rolled her eyes. "It's just a year. And I'll have my degrees when I'm done. It's kind of a work-study thing; they'll pay me enough to cover school expenses and the work credits will fulfill my senior study requirements. I've checked into it all."

Aaron's forced smile held no mirth. "And you'll never come back to us after that."

It wasn't a question, and Ginny realized with a start that he was probably right. Once she set foot on the Moon, she might not come back to Earth again, *if* it all worked out. *If* she did good work, *if* they liked her, *if* she got any job offers on the Moon after she graduated, *if* she managed to realize her dream to be a small part of *The Exodus* launch.

That was a lot of *ifs*, though, and Ginny knew it.

"It's not forever," she rasped through dry lips after too long a silence, but her father had already zizzed away into nothingness by then.

"I'm impressed, Ms. Grant; frankly, I didn't expect favorable results." Ginny's team's boss, Harold Juniper, a mousy man with no facial hair except a bushy gray moustache, regarded her from behind thick glasses as he sipped on a can of fizzy water.

"I'm glad we exceeded your expectations, Mr. Juniper. However, I would like to point out that the cryobiosis techniques our team developed only kind of imitate suspended animation, and only for a relatively short time. We can only use it reliably for a day or two. But it's promising."

"The goal has been to keep most of *The Exodus* crew asleep, while rotating only one of five sets of crewmen." Harold's glasses slipped a bit down his nose.

"Yes, but that will be practicable only if we can extend that time to months, not days. The process is expensive and rotating the crew every day or two would be cost-prohibitive, not to mention potentially dangerous. We don't know the long-term effects of repeated application."

Harold pushed his glasses back up to nose bridge. "Essentially, *The Exodus* crew would be volunteer subjects for that experiment."

Ginny's stomach turned. "I guess so." She cleared her throat. "But right now, the technology is in its infancy. It's not ready for *The Exodus* yet."

"Granted, but say it would be. Say we could keep them asleep for eighty percent of the year, thawing them out for their ten-week stint. How long would a twenty-something last on *The Exodus*?"

"It's hard to say, Mr. Juniper. We don't know at what rate they would age during such a lengthy sleep. We can't really extrapolate from the data we have; we can only speculate."

"That's good enough for me."

"It's not good enough for *me*!" Ginny protested. "I'm a scientist, not a seer."

"Give me a guess, with whatever assumptions you need."

Ginny mulled that over for a moment. "Okay, if we make the *assumption* that while asleep, aging drops to maybe twenty percent of normal—I don't think it could be zero, as much as we'd like it to be—with five teams rotating, eighty percent of the year the crew would age at that rate, making an effective aging rate over the year of thirty-six percent. Let's call it a third. Given a crewmember in his or her twenties, with an estimated remaining lifespan of sixty years—take off another twenty for child-rearing on the other end, so more like forty years—the process could add eighty years, so call it a hundred twenty years on the ship."

Harold chewed his lower lip. "So, we'd need to get up to four percent the speed of light to make Proxima b and still have time to raise the colonists from birth?" Harold sighed. "I've heard that two percent may be achievable soon, but four percent is still pretty far off."

Ginny nodded. "What about three percent? Maybe even two point five percent? I've heard rumors about one team on Lunar A looking into bio-wombs that might be close, and another on Lunar E talking about brain and spinal cord transplants into robotic bodies like they're ready to try it with humans. Is it more than rumors?"

Narrowing his eyes, Harold stared. "Where have you heard that?" He looked around guardedly and lowered his voice. "Off the record, yes. But you didn't hear it from me."

Ginny nodded. "Three percent would mean about a hundred-forty-year voyage. A hundred seventy years at two point five percent. With bio-wombs and cyborg parents raising children at the other end and—" Ginny raised a finger for emphasis "—since it's not in any of my areas of expertise, the radical *assumption* that those brains don't deteriorate like they would have in their human bodies, the crew would recover another twenty years of service, or sixty years with my team's *theoretical future* solution, for a total of a hundred eighty years."

Harold pursed his lips, deep in thought. "So, two point five percent would work."

"Begging your pardon, but why are you asking me? I'm essentially an intern, here on a lunar year abroad fellowship. Why not ask my team lead?"

"I did. Your name came up for more specific answers."

Ginny stared, taken aback. While she'd certainly done her best to help advance the team toward success, she didn't think of herself as a critical member. "I don't understand."

"I was led to believe it was your idea to study lobsters in ice?"

"Yes. It seemed a natural for cryobiosis."

"And that led to experimentation with ice inhibitors to prevent tissue damage, which allowed you to induce a hypothermic state for longer than an hour or two." Harold leaned forward. "The success of the team is largely due to your efforts, Ms. Grant. Which brings me to the other reason I asked you here today. How would you like to stay here after graduating next week?"

"I *knew* it!" Helen Grant whined, her face bloated like soft cheese left out overnight. "I *knew* you were never coming back to us. Your father should never have let you take STEM!"

"Calm down, Mom. It's not forever. I'm doing good work here, and they like me. I can't tell you what I'm working on, but if we can take it to the next stage, it will be big. Trust me."

"I don't *care* how big it is. I just want you to forget all this nonsense and come *home*. We weren't made to go to the stars."

Grimacing, Ginny muttered through gritted teeth, "It's what I *want*. Can't you just be happy for me?" She knew it wouldn't happen.

Her father zizzed into view. "You're coming home for your graduation, at least?"

Ginny sighed. "No."

Her mother burst out crying and zizzed away. Ginny couldn't see her anymore, but her sobbing continued as a continuous backbeat of sadness. Aaron looked at her. "We don't even get to see you graduate from the school that's been bleeding us dry for years?"

Ginny refused to rise to the bait. "The school has a graduation ceremony for us here on the Moon. There's a lot of fellows up here."

Aaron cringed and shook his head. "The Moon. We can't... we can't... the Moon... you know we can't." He looked away long enough to do something to his face that looked suspiciously like wiping something away, but when he turned back to his daughter, there was no evidence he'd shed a single tear.

Ginny's heart clenched. "There's a live holo-feed. I'll send you the information. It will be like you're here."

This time, she zizzed off the v-call herself, with the full knowledge that it might be the last conversation she'd ever have with her parents.

"Surprise!" Ginny wasn't surprised to see her friend sporting a scraggly brown beard, or even the mostly blond peach fuzz on his upper lip, but she was indeed surprised to see him on the other side of her apartment door on the Moon.

Stumbling backward to yank the door open, she nearly fell over. She imagined she made quite the impression, with her anime-shocked face and elephantlike grace.

Robert lunged forward and caught her before she fell, leaving her bent backward much like a dancer whose partner just dipped her. For a moment, she thought he'd kiss her, and she'd have to tell him—again—that it would never work between an Earther and a Moonie. Instead, she reached a hand up to tug playfully on his beard. "I v-called you just a couple weeks ago, and you were clean-shaven."

"Crazy, huh? Turns out I can grow a full beard in ten days. Who knew?" He righted Ginny and stepped back, looking her up and down. "You look great, Red! Your v-calls only go neck up. If I'd known, I'd have insisted you go at least waist up."

Ginny blushed, trying to ignore the underlying romantic interest in that statement. "Thanks. I liked the physical training I had to go through to get here, so I've kept it up. Why are you here? *How* are you here?"

"Virgin Galactic has shuttles, you know."

"Yeah, *expensive* shuttles."

"That look on your face when you saw me was worth a few month's wages."

Ginny frowned at him. "A few *months*?"

"Come on, I haven't seen you in five years!" He shook his head side to side in disbelief.

"We v-call every week or two!"

"Not the same, babe." Robert smiled broadly, lighting up his face.

Ginny felt a little dizzy. He really was cute, so much so that she realized she'd forgotten to chide him for using a term of endearment better suited to his girlfriend.

"Besides," he added, "I figured it would be hard to celebrate your promotion a world away. Cryonics team! That's what you've always wanted! Congrats, Red!" He moved in to hug her.

Flinching ahead of what she knew would be an awkward situation, Ginny relaxed and let him hug her. Unexpectedly, she found the embrace comforting. The hug went on too long for her comfort, though, and she squirmed away, stepping back. "Let's celebrate, then. Dinner. I'm starving. You're buying, moneybags."

They talked through dinner, and Ginny found she'd been experiencing a lack of laughter. Funny how she hadn't noticed until now. The waitstaff kept giving them significant looks, only swooping in to refill their wine glasses during slight lulls in the boisterous conversation.

"So, is *The Exodus* ever going to launch, Red? Since you came up here, it's been delayed, what, twice?" Robert cocked a questioning eyebrow, took a large bite and chewed slowly, clearly giving her time to answer.

It took her a moment to decide what to say. "That's classified, Rob. I don't want to have to kill you, so you'll have to wait to hear it on the news. How's my Mom doing?" She stuffed a large bite into her own mouth, chewing slowly. His turn.

"You know Feffers. Making trouble wherever they can."

Ginny stopped chewing. She'd seen the news; Feffers had

upped their game, actively sabotaging scientific progress whenever possible. There were reports of protester mobs preventing shuttle launches, critical research disappearing from company servers, even space-based facility bombings. Were her parents involved in any of that?

Robert swallowed quickly. "Relax; they've never been arrested, so far as I know."

Ginny said, "Good," around a mouthful of food. Classy.

"But they're not quiet about their beliefs. Your mother started a support group for what she calls 'abandoned families'—it's called Mothers Against Space Travel."

Ginny choked. "MAST?"

"Yup. And the MASTers—yes, they actually call themselves that—are growing. Big enough that the government is starting to pander to them."

"So long as we get regular supplies shuttled up here." Ginny shrugged.

Robert stared at her apparent callousness.

"What? The work we're doing up here is *important*, Rob. For *all* of us. It goes way beyond government. This is all for *humanity*. They need to get on board."

"Calm down, Red. I agree with you. But you must agree they've got a valid grievance. In your mother's eyes, your drive to go to space means she'll probably never see her grandchildren, if you ever provide any. Since you're an only child, your parents' entire genetic investment is tied up in you, and you're taking it away from Earth. You can see where they'd feel a bit betrayed."

"Betrayed? Betrayed?" Ginny's voice shrilled too much to go unnoticed. A waiter rushed over with the wine bottle, and Ginny clapped her mouth shut. After the waiter left, she lowered her voice. "If *anyone* should feel *betrayed*—"

"I know, babe."

There. He did it again. Ginny was too steamed to say anything, knowing it would come out more sharply than she intended, but she'd have to discuss it with him at some point. She took a few calming breaths. "They've never understood me, Rob."

"I know. I know."

Ginny squeezed her eyes shut. "I haven't spoken to them since graduation, you know. I just . . . can't."

Clinking his glass on hers, Robert said cheerily, "Hey, we're celebrating, remember?"

Ginny sighed and let out a long breath. "You're right. Thanks so much for coming up. It really means a lot to me. How long before you go back?"

Robert grinned. "That's the other thing we're celebrating, babe. I got a job here on the Moon."

"Did you hear? JeffGate works! *The Exodus* launches this year!"

Ginny jumped out of the way of the screaming teen in an orange unibody careening down the hallway. Smacking against the wall with bone-shaking force, she bounced back and stared after him. The Jeffrey Gate works? The ever-elusive wormhole in space actually *works*? The implication was staggering.

Stepping through the door into the cryonics lab, Ginny stopped dead in her tracks. Sally Valencia, the director of Lunar C, stood there, waiting for her.

"Let's talk." She headed directly to her office and took a seat, waiting for Ginny.

After a deep breath, Ginny strode in, closing the door behind her, and settled into her chair. "What can I do for you, Sally?"

"I imagine you've heard about the JeffGate?"

"I did."

"And you understand what it means for your department?"

Ginny cocked her head but said nothing. All the work she'd been doing for the last five years was unnecessary if the Jeffrey Gate worked. A controllable wormhole meant that *The Exodus* could make the trip in a fraction of the time that would be needed at their target of 2.5 percent of light speed. She knew they were working on the Jeffrey Gate, but like the multiple failures of variations of the Alcubierre warp drive, she didn't pay too much attention to it, since it was more fiction than science.

"I can guess."

"Well, then," Sally said simply. "You know what to do then. How many do you *need* to continue the research? I'm told you're close. The technology might still be useful."

Need? How many of the 152 people on her payroll were *needed*? Her first instinct was to blurt out, "All of them," but she bit back that retort and decided to give it a bit more thought.

Sally rose. "Get back to me this afternoon. I have several more unpleasant conversations to have this morning." She breezed out, leaving Ginny in her chair, stunned.

The rest of the morning was a bloodbath, personnel-wise. Ginny heard rumors of entire departments disappearing in the JeffGate wake; at least she still *had* a department.

If she wanted it.

She wasn't sure she did, any more. Her entire life since *The Exodus* was announced had been carefully aimed at space. She so wanted to be a part of the brand-spanking-new space culture that she'd willingly abandoned her own parents to get there.

Just like her father had said she'd do.

She'd bet her career on cryonics and lost. It seemed a safe bet back then. There were documented cases of people being revived from hypothermia-induced suspended animation. There was even a man, Mitsutaka Uchikoshi, who essentially hibernated for twenty-four *days*. It stood to reason that if it could be controlled, it could be extended to forty weeks, maybe even more.

Ginny's cryobiosis team never came close to forty weeks, but they *did* successfully replicate Uchikoshi's twenty-four days repeatedly—at least, with pigs—even pushing it to thirty days before she left for the cryonics team. Her new team was more speculative, riskier, definitely more science-fictiony, but the potential rewards were massive if they succeeded. Long-term sleeper ships could take as long as they needed to go dozens, maybe hundreds of light-years away, waking up its occupants at the end, fresh as daises, their full lives still ahead of them.

Yes, it was less likely to work than cryobiosis, but it was still far more grounded in real science than the Alcubierre warp drive or the Jeffrey Gate.

Except, apparently, it wasn't.

With a weary sigh, Ginny v-called Sally with her staff recommendations and her resignation.

"I'm glad you came, Gin." Robert's disarming smile invited her in to his apartment more than the door swinging open beside him. Garlic wafted from his kitchen, stinging her nose, which she immediately wrinkled.

"What *is* that smell?" She stepped through the door, shutting it behind her, and followed Rob to the kitchen.

"My grandmother's famous garlic chicken soup. She says it

cures anything. With a dozen bulbs—not cloves, *bulbs*—it'll cure your taste buds, guaranteed. Anything else is debatable, but it's worked for me pretty well so far. Colds, flu, herpes—"

Ginny looked up at him sharply.

"I'm kidding. I don't know about that last one."

"What about crushed dreams?"

Robert didn't answer, just gathered her into a tight hug. "I heard about the cuts. I'm so sorry. I'm surprised they'd cut you loose, to be honest. You're the Golden Girl."

Ginny relaxed into him, resting her head on his hard chest. "They didn't. I did."

Robert broke the embrace, but still held her shoulders. "Really? Why?"

Ginny sniffed, trying to hold back what she knew would be a flood of tears. How to explain to him that she felt like her entire life had been wasted? How to convey just how crushed she felt, how invalidated. "Maybe my parents were right. Maybe I should have tried to Fix Earth First."

"No," Robert said emphatically. "No, you don't." He guided her over to his sofa but didn't take his hands off her. He sat down and allowed her to nestle into him. "You're not a Feffer, never were, never will be. What you did was *right*, for you, for *all* of us. You didn't *fail*."

"Yes, I did. We never really figured out cryobiosis, much less cryonics."

"Sure you did. You brought back pigs after a month, didn't you?"

"Pigs. Not humans. Too many ethical hoops to jump through."

"I heard Lunar E had done it with humans."

Ginny glared. "You know the E means 'experimental,' right? Not 'ethics.' In fact, if you believe the stories, their supposed volunteers may not be." She sniffed again. "More likely they were being punished."

"But they still did it?"

"I'm not debating this tonight, Rob. That's not the point. They weren't my team."

"But they were using *your* research, and it *worked*, Gin! *That's* the point. Think about it. Ignore the ethics."

Ginny folded her arms over her chest, taking Robert's along for the ride. "I suppose."

"Exactly," Robert whispered into her ear. "See? You're a success, just like I always told you."

Ginny twisted around so she could see him. "You always know what to say to make me feel better." She smiled but didn't turn back around.

They stared at one another, saying nothing. Ginny heard her heartbeat pound in her ears. She saw her reflection in Robert's eyes, not a mirror image, but the way he saw her, and what she saw was beautiful, not the nerdy redhead she saw in her mirror, but a goddess incapable of wrongdoing, destined for greater things.

She closed her eyes and kissed him, hard.

"Sorry to wake you, Ms. Grant, but I couldn't wait."

Benson Mgatha's booming basso voice echoed through Robert's bedroom. His voice was well known to anyone on the Moon.

"One-way holo," she directed. The familiar chocolate skin and shocking white curls of the charismatic architect of *The Exodus* mission zizzed to life in front of her. "Sorry, sir, I'm not decent."

In fact, she was completely naked, despite instinctively gathering some of the bed sheets to cover her chest. She glanced behind her and realized Robert's bare chest would have been in view, as well.

"That's quite all right, Ms. Grant. I knew I'd be waking you. I take no offense."

"Thank you, sir."

"You're welcome. I'll come straight to the point. As you likely know, the crew roster for *The Exodus* has been filled since nearly the beginning of the project."

"Yes, sir."

"What you may not know is that occasionally substitutions must be made. It has been quite a long time, after all. Some of the crew elected not to participate after waiting so long with repeated launch rescheduling. Some are victims of accidents. You get the idea."

"Of course."

"I personally select their replacements. Replacements that impress me. Replacements like you."

Ginny blinked. "Me?"

"Yes, you. I've watched your progress since the board first extended a fellowship to you. You have impressed me."

"Me?"

The architect laughed heartily. "I was disappointed when I first heard of your resignation, but then I realized it was an opportunity that I felt would benefit us both."

Ginny opened her mouth, but only managed to croak.

"How would you like to accompany me on the maiden voyage of *The Exodus*?"

"You know I have to do this."

Ginny stared into Robert's eyes for the last time, memorizing the dark green circle around his brown irises. She ran her fingers lightly over his sandpapery cheeks, his silky hair. She breathed in his slightly tangy scent and closed her eyes, drawing close to his chest one last time.

"I know. The stars call you; they always have."

She whispered, "That's why I never wanted this to happen between us." She looked up at him. "But I'm glad it did."

"So am I," Robert choked. "So am I. I love you, Ginny. I always will; no matter where you are, no matter how far away. Remember that."

"I love you, too," she whispered.

Pulling away from his embrace was the hardest thing she ever did, but it was the right thing to do, and they both knew it. Robert could change, did change for her, she knew that. He'd never had the urge to go to space, never saw it as his destiny like she did. But he followed her to the Moon, anyway, and she knew he'd follow her to Proxima b if he could.

But he couldn't.

He needed an invitation from the architect, and that was one thing she couldn't give him.

She would go alone.

She had no problem leaving behind her estranged parents. They'd gone full-on Feffer with a vengeance, so much so that she'd stopped inquiring about them, for fear that one day the answer would be too heinous to consider.

She had no problem leaving behind her coworkers and the few friends she'd made on the Moon. They all knew she'd leave eventually, and so did she, so Ginny hadn't really formed close attachments.

Except for Robert. *That* hurt, but this was her dream, her

life, and he would never forgive her if she stayed. He knew she had to go and why, knew he had to let her go, no matter how much it hurt him. He'd carry those scars the rest of his life, but he wouldn't hold her back. He loved her too much.

"I have to go."

Ginny stared out the viewport at Proxima Centauri b, far below the orbiting spacecraft. The swirls of ice blue and deep purple painting its surface were especially striking today. She'd made the voyage in *The Exodus* well over a year ago, one of the select pioneers chosen to try to establish a colony on Proxima b.

They still hadn't done it, but they were making progress, even in the face of the fierce stellar winds plaguing its twilight surface.

They'd learned a lot about Pb—or Lead, as they'd dubbed it, as its initials were the chemical symbol—since they'd arrived nearly seven months after entering the wormhole. The Morris–Thorne prediction-model calculations turned out to be mostly correct for the size of the wormhole generated by the Jeffrey Gate, as had most of their prediction models.

All except the one Ginny held in her arms. *That* wasn't predicted. Two months into a wormhole was an exceptionally bad time to find out you're pregnant, but she wasn't about to terminate him, either.

Two months after arriving at Lead, the first human ever to be born in another solar system arrived in the universe.

Ginny named him Robert, after the father he would never know. She cried at his birth, but not from the pain of childbirth.

Now, she held him to her chest and tickled his nose.

"Look out there, Robert. That's your world. Lead. You're the first Leader. The very first. Isn't it beautiful?"

Robert cooed.

"How old is he, now?" Benson Mgatha's basso boomed behind her.

Ginny spun around. "Benson! I didn't know you'd come back from the surface yet."

"Just arrived." He paused for a few seconds, and then added off-handedly, "Little Robert is almost six months old now, isn't he?"

"Yes. My little spaceman." She tickled his chin.

The architect wandered over to get a closer look. "He looks like his father."

"He does." Ginny tickled him again and was rewarded with another coo and a yawn. It took a few moments before the words registered. "How do you know that? Robert wasn't on *Exodus* projects."

Benson Mgatha grinned.

"Maybe because he met me." Robert's voice—or at least what Ginny thought she remembered as his voice—came from the hallway.

Ginny's heart skipped several beats when Robert—her Robert, in the flesh—stepped into the room.

She ran to him, careful not to jostle her little spaceman too much and wake him now that he was fast asleep.

"How?" She looked between him and Benson. "Why?"

The architect was still grinning. "We're trying to establish a colony, remember? You just had a head start on the rest of us. You were unmistakably pregnant by the time we got here, so I sent for the child's father, so that Lead's very first colonist would have all the love and support he'd need. Morris and Thorne were right about *little* wormholes, too. Very fast, excellent for communications."

Ginny looked out at Lead, looked back at her baby sleeping in her arms, and finally at his smiling father.

She was finally home.

Afterword

Les Johnson

> 2 *Vanity of vanities, says the Preacher,*
> *vanity of vanities! All is vanity.*
> 3 *What does man gain by all the toil*
> *at which he toils under the sun?*
> 4 *A generation goes, and a generation comes,*
> *but the earth remains forever.*
> 10 *Is there a thing of which it is said,*
> *"See, this is new"?*
> *It has been already*
> *in the ages before us.*
> 11 *There is no remembrance of former things,*
> *nor will there be any remembrance*
> *of later things yet to be*
> *among those who come after.*
>
> —The Bible, Ecclesiastes 1:2–4, 10–11
>   English Standard Version (ESV)

Times change, and the individuals are different, but the universe and the human condition remain the same. At least, that's how Solomon saw things as recorded in the Bible; and, based on our understanding of human history and the best prognostications of the future in much of contemporary science fiction, most of us make this implicit assumption. We have *New York Times* bestselling novels of the far future with protagonists much like ourselves dealing with and making sense of the world and worlds newly imagined. They may be humans augmented with technology or bioengineered to have nearly infinite lifespans, but by and large, they are still just human, like you and me, encountering their world(s) and making sense of them in much the same fashion as people have done since before the Ecclesiastes was written.

Writers make this assumption because it is difficult to assume anything else. How many books have successfully described aliens that are truly alien versus those that are basically funny-looking humans, interpreting actions and situations in much the same way as you or me? It is difficult to think differently. How does one think like an alien that likely sees, hears, touches, senses and, most importantly, interprets their environment in ways very distinct and different from all other life on Earth? After all, how many sentient alien civilizations have we encountered and studied so far? (Hint: zero)

As a scientist, science fiction reader, and writer, I am constantly thinking about the future and the scientific breakthroughs that are happening all around us that will shape it. I've read science fiction stories with spaceships traveling faster than light, about future civilizations where interstellar travel is commonplace, and stories that tackle the impact of the information technology revolution in which we become human/machine hybrids. The stories that thoughtfully deal with the downstream biological, sociological, psychological, and societal impacts from these changes are few and far between.

When we imagine ourselves being on the bridge of a starship exploring new worlds, how do we envision our shipmates? In the United States fifty years ago, the culture imagined a mostly Caucasian human crew with a few people from different ethnic groups thrown in to make us feel better about our willingness to embrace a diverse human future. Just to shake things up, there was even a humanoid alien added to the crew. But he wasn't really alien; he was just someone with a humanlike personality who happened to be raised on another planet.

Today, we might envision that starship crew to be yet even more diverse, with those who were obvious minorities fifty years ago now thrust into more leadership positions, although there might still be the one-off humanoid alien to keep it interesting. In other words, we are likely to envision our starship crew as being mostly like ourselves, with our same hopes and fears, our same biological needs, and our same urgent awareness of an all-too-brief lifespan and mortality.

Who are we missing? The person with bioengineered super strength and super abilities embodied in Timothy Zahn's Cobras. The higher-IQ humans who don't waste time with that *sleep* thing like those in Nancy Kress's "Sleepless" series. Then there are those among the crew who have had their aging genes turned off, making them, barring an accident, immortal. (If you were immortal, would you risk your life on the edge of known space exploring the universe? I probably would not, and I suspect most of the species would also rather play it safe.)

Our becoming *Homo stellaris* won't be limited to biological augmentation. It is here that science fiction—and humanity—have been treading successfully for quite some time. We will correct our biological defects and augment our physical and mental abilities using computers and general mechanical technologies. We will have optical implants to give us better than 20/20 vision and aural implants to allow us to eavesdrop at will on anyone talking within half a mile of our location. We will have communication implants that will allow us to use future computers (or an evolved version of the internet) sublingually, allowing us instant access to the combined knowledge of all humanity without the need of some external device like cell phones and laptop computers. (I suspect we will, at least some of the time, use this ability to watch cat videos.) Implant-to-implant communication will enable a virtual form of mental telepathy, revolutionizing interpersonal communication. We will become cyborgs.

In fact, many of us today are, technically speaking, already cyborgs. If you wear glasses or contacts, or have had cataract surgery, then you are partly cybernetic. If you are diabetic and require an insulin pump to keep you alive, then you are a first-generation cyborg. If you have a pacemaker or artificial heart valve, then you, also, are a cyborg. The fact is that many of us are already becoming transhuman and we don't give it a second thought. Nor, I suspect, will the third or fourth generation of humans who have a computer or electromechanical implant, whether it be enhanced vision or hearing, or those communication implants mentioned earlier. To these future versions of ourselves, they will be as common and mundane as my eyeglasses are to me today.

Let's get back to our imagined starship. Would you imagine being on David Brin's ship *Streaker*, with intelligent, genetically engineered or "uplifted" dolphins and chimpanzees? If so, then how would you envision working with them? Would the aquatic dolphins be modified to survive outside of a water tank, floating in the zero gravity of deep space along with the humans and chimpanzees? Or would you imagine them being in flooded segments of the ship where you only interact with them via technology? Getting yet more radical, and equally plausible, why not imagine interacting with them in a flooded ship using *your* gills instead of your lungs to breathe? There is no reason to only modify animals to fit in human environments when we can modify humans to live and work in the environments native to other species.

In fact, with the current rate of discovery and innovation in the areas of biology and genetic engineering, it is extremely difficult to read stories of humans in the mid-term future (approximately one hundred years from now), let alone stories set more than two hundred years in the future, that don't deal with fundamental changes in human biology and society. We are soon going to fundamentally alter what it means to be human—or a dolphin—or a chimpanzee. We will be tinkering with our own genome and those of the other animals that inhabit planet Earth with us. This will inevitably alter our social, civil, and political interactions as well. How will society adapt to these changes? When reading about interstellar exploration and wars, outer solar system exploration and colonization, and stories that postulate what life will be like on Earth in that time period without genetically modified humans or cyborgs, suspension of disbelief is the only option.

We will extend our natural lifespan. Chimeras will abound: some intelligent, some not so much. Augmented humans will be our warriors and explorers. Creatures once human will live in the oceans of extrasolar planets, in the upper atmospheres of planetary gas giants, and on the surfaces of worlds much like Earth but with different atmospheric-gas constituents and ratios.

Will these be horrific worlds with repression and slavery of new sentient, but low-intelligence species of our own creation? Perhaps. Will there be worlds in which we have tinkered and eliminated most forms of disease and genetic defects? Almost certainly. Will the evolutionary path of humans on Earth be forever altered and diverge rapidly from the augmented evolution we are already rapidly experiencing? Absolutely. Will these changes affect how we interact with one another and our politics? Yes.

But, in the end, I suspect sentient beings, no matter their form, will be much like us. They will live, learn, love, search for meaning, and ultimately die.

In other words, as we venture to the stars, people *will* remain the same—only not so much...

Tennessee Valley Interstellar Workshop

This anthology was inspired by the work of the Tennessee Valley Interstellar Workshop, a group of people who believe that humanity not only *can* go to the stars one day, but naturally *must*. They, and participants from across the globe, understand that none alive today are likely to live to see that dream carried out. Nonetheless, they believe it to be a worthy-enough goal to begin the foundational work, and aim to bridge relationships between businesspeople, engineers, ethicists, politicians, military personnel, and artists so that it can become reality. The breadth of specialties here is no accident. Questions surrounding interplanetary and interstellar efforts are as numerous and varied as the people who will eventually spend their lives in the void between planets and stars. Though TVIW was originally a regional organization (viz. the American Southeast), it is now internationally recognized, with its events attracting speakers and attendees coming from all over the world.

Why does TVIW gather to discuss the challenges and opportunities of interstellar travel? What benefit do we glean from becoming a People of the Stars? We are compelled by our nature to think positively about the future of humanity in a beautiful yet extremely hostile universe. Life on Earth is wonderful and we should do what we can to protect and preserve it here, but there is more to explore. Among the billions of galaxies, stars, and planets, we sense a call to explore. That exploration cannot be haphazard or careless. It cannot unduly endanger the explorers nor the systems into which they travel. Our technological gains are contemptible if we only travel elsewhere to repeat the calamities visited on our own species in the past. That call must be careful, measured, and with the explicit intention of peaceful existence. To do this, we must push boundaries of biological, psychological, and sociological nature.

TVIW was created to foster and assist the study, research, and experimentation necessary to make human interstellar travel a reality, with untold benefits to life on Earth. TVIW Symposia are opportunities for relaxed sharing of ideas in directions that will stimulate and encourage interstellar exploration. It was at TVIW 2016 that the Homo Stellaris working track was tasked with describing the foundation of a space-based society, the possible adaptations and changes such a society would experience, and the necessary precursors to build the societal will to undertake such missions. Visit TVIW on the web, http://www.tviw.us, for more information about the organization and its activities. While you are there, fill out our contact form. We'd like to hear from you and have you join us in the goal of becoming *Homo stellaris*, the People of the Stars.

> —Joe Meany, Communications Director
> Edward "Sandy" Montgomery, President
> John Preston, President Emeritus
>
> Tennessee Valley Interstellar Workshop